COUNTER-INSURGENCY

COUNTER-INSURGENCY

Lessons from History

Edited by
Ian F. W. Beckett and John Pimlott

Pen & Sword
MILITARY

First published in Great Britain by Croom Helm in 1985

Repubished in this format in 2011 by
Pen & Sword Military
an imprint of
Pen & Sword Books Ltd
47 Church Street
Barnsley
South Yorkshire
S70 2AS

ISBN 978 1 84884 396 7

A CIP catalogue record for this book is
available from the British Library

Printed and bound in England
by CPI

Pen & Sword Books Ltd incorporates the imprints of
Pen & Sword Aviation, Pen & Sword Maritime, Pen & Sword Military,
Wharncliffe Local History, Pen & Sword Select,
Pen & Sword Military Classics and Leo Cooper,
Remember When, Seaforth Publishing and Frontline Publishing

For a complete list of Pen & Sword titles please contact
PEN & SWORD BOOKS LIMITED
47 Church Street, Barnsley, South Yorkshire, S70 2AS, England
E-mail: enquiries@pen-and-sword.co.uk
Website: www.pen-and-sword.co.uk

CONTENTS

List of Maps   vi

Introduction to Second Edition *Ian F. W. Beckett*   vii

Introduction to First Edition *Ian F. W. Beckett and*
    John Pimlott   1

1. The British Army: the Dhofar Campaign, 1970-1975
    John Pimlott   16

2. The French Army: from Indochina to Chad, 1946-1984
    John Pimlott   46

3. The American Army: the Vietnam War, 1965-1973
    Peter M. Dunn   77

4. The Latin American Experience: the Tupamaros
    Campaign in Uruguay, 1963-1973 *F. A. Godfrey* 112

5. The Portuguese Army: the Campaign in Mozambique,
    1964-1974 *Ian F. W. Beckett*   136

6. The Rhodesian Army: counter-insurgency, 1972-1979
    Ian F. W. Beckett   163

7. The South African Army: the Campaign in South
    West Africa/Namibia since 1966 *Francis Toase* 190

Notes on Contributors   222

Index   223

MAPS

1.1	Southern Dhofar	18
2.1	The French in Indochina	48
2.2	Algeria	61
2.3	Chad	69
3.1	Vietnam	78
3.2	The 'Iron Triangle'	91
4.1	Latin America	113
5.1	Mozambique 1974	137
6.1	Rhodesia 1972-9	165
7.1	South West Africa/Namibia	192
7.2	The War Zone: Namibia/Angola	193

INTRODUCTION TO SECOND EDITION

Back in 1985, when the first edition of this book was published as *Armed Forces and Modern Counter-insurgency*, there seemed little interest in counter-insurgency, at least not in Britain. There were a handful of general histories of guerrilla warfare, often of a popular rather than an academic kind. Moreover, despite the fact that guerrilla warfare had always been the most prevalent form of conflict and certainly so in the twentieth century, if not before, counter-insurgency was little studied. This was all the more surprising in the case of the British experience of warfare, the instances of conventional warfare as opposed to some form of low intensity conflict since 1945 being few. In so far as any British counter-insurgency campaign had been seriously analysed, it was the Malayan Emergency of 1948-60 that held the field, not least because of its central role in formulating the principles of counter-insurgency espoused by the influential Sir Robert Thompson.

In 1971, however, Frank Kitson had challenged the assumptions underpinning Thompson's approach. Kitson had called for British soldiers not only to consider the practical requirements of counter-insurgency but also to look beyond the Malayan example. New ideological, political and commercial imperatives encouraging intra-state conflict and insurgency were already beginning to emerge amid the breakdown of the international bipolar political system and the emergence of identity politics and of many more non-state actors. It is a trend that has continued ever since so that varying forms of low intensity conflict are even more familiar now than in the mid 1980s.

It was in the spirit of Kitson's challenge that the late John Pimlott and myself introduced counter-insurgency as a special subject for study at the Royal Military Academy, Sandhurst in the early 1980s. In the fifteen years that I taught war studies at Sandhurst between 1979 and 1994, the syllabus changed with bewildering rapidity and the course lasted only a few years. It gave rise, however, to *Armed Forces and Modern Counter-insurgency*, the intention being to arrive at a general framework for analysis that could then be applied to a variety of different campaigns. Ronald Haycock had edited a rather discursive collection of essays, *Regular Armies and Insurgency*, in 1979. It seemed to John and myself that, while this collection was similarly organised as a series of case studies, it did not require individual authors to try to conform to such a analytical framework as we envisaged. Subsequently, the case study approach has remained popular as in David Charters and Maurice Tugwell's edited collection, *Armies in Low-intensity Conflict: A Comparative Analysis* (1989) and Daniel Marston and Carter Malkasian's *Counterinsurgency in Modern Warfare* (2008). Subsequent to *Armed Forces and Modern Counter-insurgency*, a similar methodology of analysis was used in my edited collection, to which John contributed, *The Roots of Counter-insurgency: Armies and Guerrilla Warfare, 1900-45* (1988)

Inevitably, as in the latter case, work undertaken since 1985 has both modified the analysis of some campaigns we explored but also extended knowledge of other counter-insurgency campaigns such as the 'Troubles' in Northern Ireland, which continued until 1997. Equally, of course, the South African experience analysed by Francis Toase in 1985 changed substantially with the agreement on majority rule in 1994. Ironically, many former insurgent movements once in power were themselves faced by insurgency, as suggested by the essay collection, *The Counter-insurgent State*, edited in 1997 by Paul Rich and Richard Stubbs.

In many respects, the model framework suggested by *Armed Forces and Modern Counter-insurgency* remains just as valid now as in 1985. Rather than Thompson's classic 'five principles' the model suggested was a six-fold one. Following recognition of the political nature of insurgency, the second basic requirement of a successful counter-insurgency strategy appeared the recognition of the need to ensure co-ordination of the military and civil response; a third, the need to ensure co-ordination of intelligence. A fourth requirement was the separation of the insurgents from their base of popular support either by physical means or by a government campaign designed to win the allegiance of the population. A fifth requirement was the appropriate use of military force against those insurgents separated from the population. Finally, long-term reform addressing those political and socio-economic grievances that have contributed to the insurgency was necessary in order to ensure that it did not recur. Examining each of the case studies in these terms was the key to an understanding of why some had succeeded and others failed.

Much of this is now widely accepted. Whatever the degree of support for it, insurgency can only be realistically tackled through a primarily political response. Where initial support for insurgency amongst an uncommitted population depends upon the exploitation of particular grievances then attention to those grievances on the part of government may perhaps be more significant than the perception of winning and losing. Clearly, government needs to project a viable political and socio-economic programme that offers significantly more future promise than that of the insurgents. Yet, the ability of government to address fundamental grievances may be part and parcel of that wider perception of winning and losing. The population's overall sense of security is the key whatever other tangible rewards may be on offer from one side or the other and, indeed, without security it may not be possible to deliver other rewards.

To win a counter-insurgency campaign, therefore, implies establishing sufficient legitimacy to incorporate a critical mass of the population within the government camp. Establishing or re-establishing legitimacy will also imply promoting a sense of credible security as a basis for governance. In essence, this is a matter of 'winning hearts and minds'. It has been suggested on occasions that 'if you have them by the balls, their hearts and minds will follow', but, while this may convey an element of truth, the reality is that victory will go to those who are best able to meet both security concerns and basic aspirations. Most models of counter-insurgency, therefore, recognise the imperative of putting into place sufficient reforms and mechanisms to ensure that the insurgent challenge will be neutralised and will not recur.

Other models have been suggested since 1985, all of which begin from a position of the overarching objective of establishing or re-establishing legitimacy and with the necessity for the political nature of the response to be paramount. The 'Max Factors' evolved by Max Manwaring and John Fishel suggest seven dimensions in which success or failure is determined, namely host government military actions, action against subversion, unity of effort, the retention of external support for host governments, the struggle for legitimacy, the separation of insurgents from the population, and the reduction of external support for the insurgents. Anthony Joes has suggested a two-fold approach in shaping the strategic environment and isolating the insurgents. Shaping the environment requires redressing legitimate grievances, deploying sufficient troops to restore order, and excluding external support. Separation of insurgent from the population requires adherence to the rule of law co-ordination of intelligence, food and weapons control, and the use of amnesties.

Interest in counter-insurgency has grown appreciably since 1985 as suggested by such alternative models. Further stimulus has derived from events in Iraq and Afghanistan since 2001. Scholars such as Susan Carruthers, David Charters, John Coates, Raffi Gregorian, Randall Heather, Robert Holland, Keith Jeffery, Tim Jones, David Percox, Richard Popplewell, Anthony Stockwell, and Richard Stubbs have extended knowledge just in terms of the British experience. Thomas Mockaitis, John Newsinger, and Charles Townshend have provided new overviews of British operations. Their work is reflected in my own *Modern Insurgencies and Counter-insurgencies: Guerrillas and Their Opponents since 1750* (2001), while some of their key journal articles were reproduced in my edited selection, *Modern Counter-insurgency* (2007).

One aspect of British counter-insurgency, for example, that has aroused particular controversy is the issue of 'minimum force'. Initially, it was suggested by John Newsinger's questioning of the concept in the context of the Mau Mau insurgency in Kenya between 1952 and 1959. Newsinger and Mockaitis debated the issue in the journal, *Small Wars and Insurgencies* in 1992. In passing, it might be noted that *Small Wars and Insurgencies* itself, first published in 1990, was a real indication of the growing academic interest in the subject. Subsequently, books by David Anderson and Caroline Elkins in 2005 revived the controversy, and there has been further critical appraisal of the notion of the exceptionalism of British minimum force in the inter-war period by scholars such as Matthew Hughes and Huw Bennett.

Equally, knowledge of the French experience, particularly in Algeria, has been greatly extended by scholars such as Martin

Alexander, Tony Clayton, and John Keiger. There has been less on Portuguese, Rhodesian and South African campaigns though J K Cilliers's work, available only as a thesis while *Armed Forces and Modern Counter-insurgency* was being prepared, was published later that year. Moorcraft and McLaughlin's *Chimurenga* (1982), was republished in an updated version under a new title by Pen & Sword in 2008. Peter Godwin and Ian Hancock's *Rhodesians Never Die* was published in 1993 and John Cann's important analysis of Portuguese counter-insurgency appeared in 1997 and his edited collection, *Memories of Portugal's African Wars* in 1998. Of course, the Vietnam War has continued to generate an enormous amount of new scholarship, so much so that it would be impossible to encompass it all here. Representative perhaps is the work of scholars such as Eric Bergerud, Larry Cable, Andrew Krepinivich, Michael Hennessy, Richard Hunt, John Nagl, and D M Shafer.

Notwithstanding this body of work, *Armed Forces and Modern Counter-insurgency* retains its value. Indeed, the influence of the late John Pimlott, who continued to teach at Sandhurst until his early and tragic death in 1997, can still be discerned in current British counter-insurgency doctrine. Bearing in mind the six-fold framework outlined in *Armed Forces and Modern Counter-insurgency*, it can be noted that they bear a striking similarity to the six principles enunciated in *The Army Field Manual Volume V: Operations Other than War* and *Counter Insurgency Operations (Strategic and Operational Guidelines)*, both issued in 1995. These six principles were the need to ensure political primacy and a clear political aim, to build a co-ordinated government machinery, to develop intelligence and information, to separate the insurgent from his support, to neutralise the insurgent, and to plan for the longer term. The latter manual was re-issued in 2001 and revised and updated in 2007. The six principles were then replaced by seven principles in *Countering Insurgency*, issued in 2008: the need to ensure political primacy and a clear political aim, to gain and secure the consent of the people, to build a co-ordinated government machinery, to effect communication with the people, to provide focused intelligence, to neutralise the insurgent, and to plan for the longer term.

IFWB, May 2010

INTRODUCTION TO FIRST EDITION

Insurgency is certainly the most common and arguably the most subtle form of modern conflict. Indeed the majority of Western armed forces studied in this volume have had comparatively little experience of more conventional kinds of warfare since 1945. Whereas irregular or guerrilla warfare prior to the Second World War represented primarily a tactical method, modern insurgency has implied a politico-military campaign waged by guerrillas with the object of overthrowing the government of a state. In essence, political, social, economic and psychological elements have been added to irregular military tactics with revolutionary intent. Thus, to a large extent, the evolution of modern insurgency has been inextricably interwoven with the emergence of communist strategies for revolution – above all, the Maoist theory of revolutionary warfare. But if insurgency is modern in terms of its ideological basis, the tactics of the guerrilla are as old as warfare itself. Similarly, what might be termed counter-guerrilla tactics are clearly just as old but, while the origins of modern revolutionary guerrilla warfare theory could be said to lie in the late eighteenth or early nineteenth centuries, modern counter-insurgency doctrine is much more a product of the late nineteenth century.

To a great extent, the disparity in doctrinal development derives from the fact that those Europeans who increasingly wrote of or alluded to guerrilla warfare (or 'partisan warfare' as it tended to be called), did so from the perspective of the guerrilla and not his opponent. Greater importance was attached to conventional operations in Europe itself where partisans were seen as an adjunct to regular armies and, indeed, often consisted of regulars detached from a main army. Thus Karl von Clausewitz, although recognising the political significance of 'people's war', wrote essentially of the role of the partisan in conventional warfare. In doing so, he drew much of his inspiration from accounts of irregular warfare such as those by Andreas Emmerich, Johann von Ewald and George Wilhelm von Valentini who had experienced irregular warfare either in the American War of Independence (1775-83) or the early French Revolutionary War. Clausewitz was also, of course, aware of the development of guerrilla war in La Vendée (1793), Spain (1808), the Tyrol (1809) and Russia (1812). He wrote mainly of conventional war in his major work, *On War*, but those authors who more specifically devoted themselves to irregular warfare in the early nineteenth century, such as the Frenchman Le Miere de Corvey, the Prussian Carl von Decker, the Russian Denis Davidov and the Poles Wojciech Chrzanowski and Karol Bogunio Stolzman, also examined the value of partisans in the context of conventional warfare.[1]

In so far as irregular warfare developed within Europe during the mid and late nineteenth century, it was largely in the form of brief urban insurrections. The main exception was the Franco-Prussian War (1870–1) where *francs tireurs* harassed the German armies who had speedily dispatched their conventional French opponents in the manner of the extreme brevity of nineteenth-century European wars between major powers. Where European armies did experience increasing irregular warfare was in the expanding colonial empires and it was there that a more coherent counter-guerrilla doctrine originated. The problem was that the kind of enemies encountered and usually (but not invariably) defeated by the Europeans, were widely diverse in characteristics and methods. It is perhaps an illustration of the diversity of circumstances encountered that the British Army had no standard manual until the publication of C.E. Callwell's celebrated *Small Wars* in 1896.[2] Callwell divided potential enemies into no less than six categories — European-trained armies such as the Sikhs against whom the British fought in the 1840s or the Army's own native sepoys in India in 1857; semi-organised troops such as the Afghan Army encountered in the 2nd Afghan War (1878–80); disciplined armies with primitive weapons such as the Zulu or Matabele; fanatics such as the Dervishes; true guerrillas such as the Maoris, Kaffirs or the *dacoits* of Burma; and, in a category of their own, the Boers who differed from the other guerrillas in being both white and mounted. Many of Callwell's operational principles, such as the use of rallying squares, were outdated even in contemporary European warfare, but others, such as the stress on the importance of intelligence, lost none of their relevance to 'small wars' in the following century.

Yet it seems unlikely that Callwell's book represents much more than a synthesis of individual commanders' approaches to specific campaigns of the past. Certainly the later editions of 1899 and especially 1906 merely mentioned more recent examples such as the South African experience (1899–1902). The latter war saw the development of a fairly sophisticated pacification strategy against the Boer commandos after the surrender of the main Boer army at Paardeberg in February 1900. In the next two years British columns, as mobile as the Boers themselves, constantly harried the commandos, whose room for manoeuvre was increasingly restricted by a liberal use of wire and blockhouses across the *veldt*. The Boers' support was also totally eroded by the incarceration of their women and children in 'concentration camps' and the systematic destruction of their farms, crops and livestock.[3] Such a strategy had been evolved in response to specific local circumstances, an illustration of the importance of flexibility in British colonial campaigning rather than of any particular doctrinal adherence.

Similarly, the campaigns of the United States Army against Indians or, latterly, against Filipino guerrillas were too diverse to suggest the need for a coherent doctrine of counter-insurgency.[4] In reality a fairly similar response developed to the different military threats posed by Indians, Mexicans or Filipinos, while the US Army's operations against irregulars continued to be governed throughout the latter nineteenth century by General Order 100 of April 1863, which sought to lay down a basic code on the treatment of irregulars in warfare. Nevertheless, there was always tension between the conflicting demands for humanity and severity with regard to a hostile subject population in the American experience. Indeed, it cannot be said that any of the European armies of the late nineteenth century had any real concern for what a later generation would call 'winning hearts and minds'. The Imperial German Army in particular had a notoriously casual regard for the niceties of the admittedly ill-defined laws of war, insisting on 'military necessity' and the 'rights' of invaders. Thus *francs tireurs* were shot out of hand during the Franco-Prussian War while the unfortunate Herreros and Hottentots of German South West Africa were ruthlessly slaughtered when they rose in revolt (1904–7), although the latter campaign also illustrated the difficulty of troops trained for conventional warfare adapting to anti-guerrilla tactics and the vulnerability of troops tied to fixed railway supply lines in difficult terrain occupied by the guerrillas.[5]

The one exception to the primarily military approach to counter-insurgency was that of the French. In the 1840s Marshal Bugeaud had made use of flying columns and *razzia* or punishment raids in Algeria, but in the later nineteenth century French colonial soldiers such as Joseph Gallieni and Herbert Lyautey conceived an approach to insurgency remarkably reminiscent of twentieth-century methods. Thus in French Indochina, Madagascar and Morocco the French substituted slow methodical expansion of administration hand-in-hand with military presence for Bugeaud's rapid thrusts through insurgent territory. Progressive pacification was likened to the spread of an oil slick — *tache d'huile* — involving the systematic reorganisation of the local population, who would be attracted to the administration by the range of facilities now afforded them. Lyautey's ideas in particular had later echoes in the *guerre révolutionnaire* of the mid-1950s in their implications for society and government in metropolitan France. Lyautey had, of course, been virtually exiled to the colonies following the publication of his celebrated article on the social role of the French officer in 1891. A subsequent article on the colonial role of the French Army in 1900 equally implied that the Army might regenerate French society itself, an elite emerging from colonial operations 'which tests and proves itself in military service before leading the nation to new grandeur'.[6] From this it was but a short step to the military conspiracies of the inter-war years and of the late 1950s/early 1960s, the latter strongly motivated by the development of *guerre révolutionnaire* as a counter-insurgency doctrine.

The parallels between French theory and practice in the nineteenth century and that of the later twentieth century are an indication of how long armies persisted in utilising doctrines or ideas that had worked in the past. The *tache d'huile* strategy was essentially that used by the French in Indochina between 1946 and 1954. Rather similarly, the basic British approach to insurgency, regarding it primarily as a case of the Army simply assisting the civil power in traditional colonial policing, was a constant throughout the inter-war years. Thus the two basic British texts on counter-insurgency published between the world wars — Sir Charles Gwynn's *Imperial Policing* (1934) and H.J. Stimson's *British Rule, and Rebellion* (1937) — revealed little understanding of significant signposts for the future that had already occurred.[7] Stimson, basing most of his theories on the

Arab Revolt in Palestine (1936–7), was perhaps perverse in his emphasis on the benefits of martial law, whereas Gwynn — the basic text at Staff College — drew from a variety of minor revolts, ranging from Amritsar in 1919 to Cyprus in 1931, lessons of minimum force, firm action and civilian control coupled with co-ordination of military and civil efforts. Gwynn recognised that the weapon of propaganda invariably lay in the hands of the insurgent but he totally ignored, as did Stimson, the example of the development of a politically-motivated insurgency in Ireland between 1919 and 1921. When the British once more encountered what was to become the established pattern of post-war insurgency in Palestine between 1945 and 1948, their neglect of the lessons of Ireland left them little prepared to respond satisfactorily.[8]

Thus the British were required to extemporise a response to insurgency in Palestine as it developed, rather as the German Army had only gradually evolved its response to the development of partisan warfare inside the Soviet Union after June 1941.[9] Traditional colonial policing was simply no longer appropriate in meeting insurgency that was now firmly based on revolutionary political ideology, which often eschewed direct military action against the Security Forces for political indoctrination among the population. The kind of colonial revolts experienced by the major armies prior to the Second World War, such as the Arab Revolt which the British mistakenly equated with the new Jewish insurgency developing in Palestine after the war, rarely approached the degree of fusion of political and military activity that characterised revolutionary guerrilla warfare and insurgency after 1945. Few if any of the practitioners of guerrilla warfare in the nineteenth century or before could claim the strategy for revolution outlined by the major theorists of the twentieth century such as Mao Tse-tung, Vo Nguyen Giap, Che Guevara or Carlos Marighela.

Such insurgency demanded new methods and new understanding, and a significant feature of much post-1945 counter-insurgency theory was the amount of space given over to understanding the nature of insurgency as an essential preliminary to its eradication. Thus the best known British theorist, Sir Robert Thompson, in his seminal work *Defeating Communist Insurgency* (1966), devoted approximately a third of the book to the theories of communist insurgency rather than to his own theories of counter-insurgency. Similarly, the Americans John S. Pushtay and John J. McCuen were intent on establishing the pattern of communist-inspired insurgency to which an appropriate response might be formulated, McCuen in particular developing a theory of 'counter-revolution' which precisely mirrored his analysis of revolutionary war. In keeping with the need to comprehend revolutionary method, the US Army and other branches of the US armed forces 'bought in its entirety' the first edition of Robert Taber's *The War of the Flea* when it was published in 1965.[10] In more extreme fashion the French exponents of *guerre révolutionnaire* such as Colonel Charles Lacheroy consciously imbibed communist ideology in order to defeat it. In the case of the French theorist Roger Trinquier, it could be argued that he actually succumbed to communism as illustrated by his comment that, if forced to choose between communism and capitalism, he would choose communism.[11]

Guerre révolutionnaire is also an example of a factor equally applicable to guerrilla warfare theory as to counter-insurgency: that a theory formulated out of a particular situation or experience might not be readily translated into one of universal applicability. *Guerre révolutionnaire* was shaped by the failures of the French in Indochina. Rather similarly, much British and American doctrine was specifically developed out of the early campaigns in Malaya and the Philippines respectively. The Malayan Emergency (1948–60) prompted the British further to test those elements of their gradually evolved counter-insurgency theory that had worked so effectively against the Malayan Communist Party. It was, of course, from Malaya that Thompson evolved his 'five principles' of a need for a clear political aim on the part of the government, the need to adhere to the rule of law, the need for a co-ordinated plan, the need to establish secure base areas, and the need to concentrate initially on destroying the political infrastructure of the insurgents. The keynote of the British approach, however, has been flexibility. Whereas other countries, as this volume indicates only too clearly, have slavishly followed the fashion established in Malaya of separating population from guerrilla by means of resettlement, the British themselves have recognised that individual societies are not necessarily socially or economically 'right' for resettlement.[12]

In the campaign against the communist Hukbalahap (1946–54), the Filipino Army, assisted by United States specialists, learned to moderate an earlier 'mailed fist' policy and, with the inspiration provided by Ramon Magsaysay, first as Secretary for National Defense and subsequently as President, initiated a policy of 'all out friendship or all out fist'. The despised Philippine Constabulary was fully integrated into the Army, which itself received better pay and rations to discourage looting, while some of its former practices such as reconnaissance by fire were actively discouraged. Self-contained and self-supporting Battalion Combat Teams were trained by American advisers and conducted small unit operations rather than the large-scale sweeps of the recent past, with the emphasis on subtlety rather than predictability. While the Huks were vigorously pursued into remote areas, Army engineers staffed the Economic Development Corps, building new roads, bridges, canals, schools, clinics and 'liberty wells'. The socio-economic programme effectively eroded popular support for the Huks.[13] The Philippines experience was enormously influential for American theorists such as McCuen, Pushtay and David Galula and can be seen behind much of the global counter-insurgency doctrine adopted by the Kennedy administration in 1961.[14] Still other authors on counter-insurgency such as the Australian, J.L.S. Girling, were able to draw on both the Philippines and Malaya.[15]

The most surprising aspect of this early American experience, the fruits of which largely materialised in Latin America in the 1960s, was the apparent failure to apply the lessons where the United States was directly involved in conflict against communism as opposed to merely advising others to resist it. Indeed, it has been argued that the US Army never seriously attempted counter-insurgency in Vietnam, its lack of flexibility being summed up in the memorable remark attributed to one American general: 'I will be damned if I will permit the US Army, its institutions, its doctrine, and its traditions to be destroyed just to win this lousy war.'[16]

The difference between the United States' approach to Vietnam and the British Army's approach to counter-insurgency in its numerous campaigns since 1945 belies the fact that general principles have emerged which are of near universal applicability in dealing with insurgency. Just as modern insurgency itself can be divided fairly conveniently into political, military, socio-economic, ideological, psychological and international dimensions[17] so, too, can counter-insurgency be discussed and analysed in its political/international, military, and socio-economic/ psychological dimensions. Such an approach, as followed in the individual essays in this present volume, thus enables conclusions to be drawn by way of comparison between different armed forces' experiences.

Since most modern counter-insurgency is far from being a purely military problem, it is necessary for the Security Forces to formulate a response that takes this clearly into account. Thus it has been said that the French campaign in Algeria was really seven separate conflicts being fought simultaneously, of which only one was the 'fighting war' itself. The others were primarily political struggles between and within the French, Algerian and settler communities which raged not only in Algeria and metropolitan France but across the wider global political stage.[18] Such complications have invariably required a well co-ordinated command and control system such as that adopted by the British in Malaya and Cyprus where Field Marshals Templer and Harding respectively enjoyed virtually pro-consular powers. Such co-ordination of both the civil and military effort must occur at all levels and embrace the provision of intelligence which is vital in the prosecution of effective counter-insurgency. Invariably, the British practice has been to entrust the gathering of intelligence, at least initially, to the Police and its Special Branch but in some British compaigns as well as in Rhodesia and the Portuguese colonies in Africa this has frequently resulted in friction between Police and Army. It is significant that the most recent British theorist of counter-insurgency, Frank Kitson, has expressed the belief that it is better to expand the Army than the Police in meeting insurgency in which the Police are often prime targets for terrorism.[19] It has also been frequently argued that a Police Force cannot provide the kind of operational intelligence required to enable the Army to locate and eliminate the guerrilla. The question of Police primacy is clearly one that must be resolved sooner or later in most counter-insurgency operations, but it is probably immaterial which emerges in control provided that there is agreement and co-ordination between the two.

Another aspect of counter-insurgency which has recurring significance for armed forces is the level of force to be deployed. On occasions, as in Rhodesia, the amount of manpower available may be limited by circumstances beyond the control of either armed forces or politicians. More often, however, the level of force deployed results from political decisions. Thus American manpower levels in Vietnam were always closely controlled by successive administrations intent on minimising casualties and political opposition at home, while it is logical to argue that present Soviet strategy in Afghanistan is strongly reminiscent of the limited 'enclave strategy' to which the United States was all too briefly attracted in Vietnam in the summer of 1965.[20] It must be recognised, of course, that total numbers may not relate at all to those actually available for ground operations. Thus only 22 per cent of the US forces in Vietnam were in 'teeth' arms, while an apparent British superiority over the guerrillas in Malaya in 1952 of 15:1 represented a ratio of only 2.5:1 on the ground. Of course, the British emphasis on minimum numbers and, as a consequence, 'minimum force' also reflects lack of a large reserve as well as preference. This is not the way of other states, such as the Soviet Union in Afghanistan, the United States in Vietnam or the Iraqis in their long struggle against the Kurds (1961–75), which have sought to offset lack of sufficient numbers of men on the ground by massive reliance on firepower and aerial bombardment. Rather similarly, the British have been reluctant to cross international frontiers, beyond which most successful guerrilla groups since 1945 have taken refuge, while other states such as Israel, Rhodesia and South Africa have shown no such compunction. Indeed, based on Israeli experience, the South Africans have not only undertaken 'hot pursuit' operations but also pre-emptive strikes against guerrilla concentrations in host states. The Portuguese and the French, like the British, adopted a largely self-imposed political restraint when faced with external guerrilla refuges.

The majority of counter-insurgency campaigns have begun with guerrilla action against armed forces trained and equipped primarily for conventional roles and it is largely a matter of how fast such conventional forces can be persuaded to adopt a more flexible response. In the case of the Soviet invasion of Afghanistan, for example, it is apparent that the Soviets have been extremely slow to relearn lessons of mountain warfare which were well known to the British in their days on the North West Frontier of India, despite the fact that the Soviets have experienced such insurgency against Moslem guerrillas in the Basmachi Revolt of the early 1920s. Indeed, Marshal Semyon Tukhachevski, later executed in Stalin's purges, has been described as one of the founders of modern counter-insurgency theory.[21] As yet no coherent modern Soviet doctrine appears to have emerged from the Afghan experience.[22] It is for this reason, as well as the fact that most modern insurgency is communist-inspired and aimed specifically at Western governments or pro-Western governments, that this volume is concerned with the emergence of Western counter-insurgency theory since 1945. However, as *guerre révolutionnaire* proved, the Soviets would be well advised to avoid a rigid doctrinal approach to counter-insurgency. Clearly, too, the actual military tactics employed must be related to the terrain and nature of the guerrilla threat although, again, much may depend upon the manpower available. In the case of Rhodesia, lack of numbers led to the development of the 'Fire Force' concept to offset lack of resources through a concentration on firepower and mobility. Firepower, however, may not be an adequate substitute for what Sir Robert Thompson referred to as 'same element theory' or, put more simply, having men on the ground hunting down the insurgent in his own environment. Of course, the development of the helicopter and its increasing deployment with Security Forces in the 1950s and especially the 1960s has made an enormous difference to the mobility of counter-insurgency forces. In Malaya ten minutes by helicopter was calculated as the equivalent of ten hours by foot, but experience has shown that the tendency to become 'heli-bound' in the way that the French were road-bound in Indochina must be resisted. There is little doubt that, in Vietnam, US forces often forgot that it is necessary to get out of the helicopter in order to fight effectively against the insurgent on the ground.

The reliance on technology in counter-insurgency, although not entirely misplaced (notably in the increasing use of computers in intelligence analysis and of sophisticated surveillance devices), inevitably carries the risk that it will dominate the conduct of the campaign. Generally speaking the less sophisticated the army, the better able it has been to defeat insurgency. Indeed, reliance on the helicopter or firepower may be regarded as pacification rather than counter-insurgency. The latter also requires contact with the people on the ground, for the civilian population will always be the arbiters of success or failure for both Security Forces and insurgents. The outcome, as far as the Security Forces are concerned, depends upon the security of the civilian population and their separation from the insurgent, this being an absolute requisite for the essential task of 'winning hearts and minds'. It has been customary for Security Forces since 1945 to ensure such separation either by erecting a physical barrier against guerrilla infiltration in the manner of the French 'Morice Line' in Algeria, the American 'McNamara Line' in Vietnam and the Rhodesian and South African *cordon sanitaire* of border minefields, or by resettlement of the civilian population. Resettlement was, of course, one of the great successes of the Malayan campaign but elsewhere strategic hamlets, protected villages, *aldeamentos*, or whatever the new locations for settlement have been called, have invariably failed.[23] This is not necessarily a reflection on the actual concept but upon its execution, and it is unlikely that a future resettlement would be absolutely doomed to failure if carried out with sufficient care and preliminary preparation. The winning of hearts and minds itself may take many forms, from the Rural Industrial Development Agency of the Malayan Emergency to the Civil Action Teams in Dhofar. It must, however, offer genuine opportunities for the indigenous population and genuine improvement in the quality of life. Not all Security Forces have learned this basic lesson, notably the Soviets who employ in Afghanistan a policy of 'scorched earth and migratory genocide'.[24] Indeed, for insurgency to develop at all is frequently an indication that the government of a country has failed to accommodate the basic aspirations of its population. It simply does not hold that 'if you have them by the balls, their hearts and minds will follow'. Nevertheless, it is always likely that the Security Forces will have some measure of popular support, as occurred frequently in Latin American states where armies were far more popular than police forces, and a feature of counter-insurgency since 1945 has been the degree to which all Security Forces have employed local forces. Thus only 175,000 out of the 500,000 French troops in Indochina were actually French, while the British have employed local forces as varied as the Seroi Pra'ak in Malaya and the firqat in Oman. Large parts of the Rhodesian and Portuguese Security Forces were black rather than white, while the South African Army has utilised its 'Buffalo Battalion' as well as the non-white South West Africa Territory Force in its campaigns in South West Africa/Namibia and Angola. Such local forces are invaluable in terms of their local knowledge and have frequently embraced surrendered guerrillas or those 'turned' after capture. The latter have increasingly been used as 'pseudo-forces' to penetrate and disrupt guerrilla organisation, from the counter-gangs employed during the Kenyan Emergency of the 1950s to Rhodesia's Selous Scouts.

Involvement in counter-insurgency, however, may have repercussions on armed forces. For one thing, it interferes with normal training for conventional war, as it did in the case of the British 6th Airborne Division in Palestine, which was deployed to counter-insurgency in 1946 instead of undertaking regular parachute training as the 'Imperial Reserve'. On the other hand, as the Soviets have discovered in Afghanistan, it may also prove a welcome training experience in the absence of any relevant conventional experience since 1945. Counter-insurgency by its very nature may also prove a great strain, the Security Forces being placed in a position where the slightest retaliation will do much for the opponent's propaganda machine. Thus in Cuba between 1956 and 1959, while Batista's Army invariably executed captured *Fidelistas*, Fidel Castro's guerrillas treated captured regulars well as a deliberate policy. There is little doubt that the role of the French 10th Colonial Parachute Division in the 'Battle of Algiers' in 1957 did little for France's international prestige, while something of a war crimes 'industry' developed in connection with the US role in Vietnam and dogged the Rhodesian, Portuguese and South African forces' counter-insurgency campaigns in southern Africa. Some armed forces, of course, take more notice of morality than others, Latin American states having ruthlessly suppressed insurgency along with civil liberties throughout most of the South American continent. The involvement of the Latin American armed forces in counter-insurgency also brings to mind another effect of counter-insurgency, which may be the actual politicisation of the army itself. Thus the Portuguese armed forces were partially provoked to overthrow the Portuguese government in April 1974 as a result of the campaigns in the African colonies, while the French Army's politicisation in Algeria is well known. To some extent such politicisation may result from sheer frustration with the protracted nature of modern insurgency in which results in the military field may not be relevant to the political outcome. Still other repercussions could take the form of a collapse of morale in the Security Forces as occurred in Vietnam.

Vietnam conjures up the considerable attention that has been paid since 1945 to the success or apparent success of insurgency from the Chinese Civil War onwards, although it should be noted that both the war in China and that in Vietnam ended as conventional conflicts — Maoist theory, of course, differs from some other revolutionary guerrilla warfare concepts in portraying insurgency as a means to an end, which is the creation of a conventional army capable of waging and winning conventional wars. But counter-insurgency is not always destined to fail, as witness its successes in Greece, the Philippines, Malaya, Kenya, Borneo, Oman and much of Latin America. Such success has not been easy, cheap or rapid but it has been achieved. There is clearly no universal blueprint for success in what is a most difficult form of conflict to counter, but certain principles have emerged of general applicability. It must be said that most of those principles have been a product of British experience and it is therefore no surprise that the British Army remains the most successful and practised exponent of counter-insurgency since 1945.

Department of War Studies   Ian F.W. Beckett
and International Affairs,   John Pimlott
RMAS. June 1984

Notes

1. The best account of guerrilla warfare theory through the ages is to be found in Walter Laqueur, *Guerrilla: A Historical and Critical Study* (Weidenfeld and Nicolson, London, 1977) although the more concise Lewis H. Gann, *Guerrillas in History* (Hoover Institution Press, Stanford, 1971) is also useful. By far the best account of modern insurgency is John Baylis, 'Revolutionary Warfare' in John Baylis, Ken Booth, John Garnett and Phil Williams (eds), *Contemporary Strategy* (Croom Helm, London, 1975), pp. 132–51. For Clausewitz, see the edition of *On War* edited by Michael Howard and Peter Paret (Princeton University Press, 1976).

2. C.E. Callwell, *Small Wars: Their Principles and Practise* (HMSO, London, 1896). The diversity of Victorian military experience can be traced in Brian Bond (ed.), *Victorian Military Campaigns* (Hutchinson, London, 1967), while a general survey of nineteenth-century colonial campaigning and its relevance to European armies can be found in Hew Strachan, *European Armies and the Conduct of War* (Allen and Unwin, London, 1983).

3. The most concise account of British strategy in South Africa is Howard Bailes, 'The Military Aspects of the War' in P. Warwick (ed.), *The South African War* (Longman, London, 1980), pp. 65–102.

14 Introduction

4. See J.M. Gates, 'Indians and Insurrectos: The US Army's Experience with Insurgency', *Parameters* XIII/1, 1983, pp. 59–68. For studies of specific American campaigns, see J.M. Gates, *Schoolbooks and Krags: The United States Army in the Philippines, 1898–1902* (Greenwood, Westport, 1973); R.M. Utley, *Frontier Regulars* (Macmillan, New York, 1973); J.P. Tate (ed.), *The American Military on the Frontier* (US Air Force, Washington, 1978).

5. J.M. Bridgman, *The Revolt of the Herreros* (University of California Press, Berkeley, 1981); Geoffrey Best, *Humanity in Warfare* (Weidenfeld and Nicolson, London, 1980), pp. 166–99.

6. J. Gottman, 'Bugeaud, Gallieni, Lyautey: The Development of French Colonial Warfare' in E.M. Earle (ed.), *Makers of Modern Strategy* (Princeton University Press, 8th edn, 1966), pp. 234–59; Douglas Porch, *The Conquest of Morocco* (Knopf, New York, 1983); Peter Paret, *French Revolutionary Warfare from Indochina to Algeria* (Pall Mall Press, London, 1964), p. 108.

7. Sir Charles Gwynn, *Imperial Policing* (Macmillan, London, 1934); H.J. Stimson, *British Rule, and Rebellion* (Blackwood, Edinburgh and London, 1937).

8. D.A. Charters, 'Insurgency and Counter-Insurgency in Palestine, 1945–47', (Unpublished PhD thesis, London, 1980). On Ireland, see Charles Townshend, *The British Campaign in Ireland, 1919–21* (Oxford University Press, 1975) and T. Bowden, *The Breakdown of Public Security* (Sage, London, 1977), which also studies the Arab Revolt.

9. Keith Simpson, 'The German Experience of Rear Area Security on the Eastern Front, 1941–45', *Journal of the Royal United Services Institute* 121/4, 1976, pp. 39–46; C.P. von Luttichau, *Guerrilla and Counter-guerrilla Warfare in Russia during World War Two* (Department of the Army, Washington, 1963).

10. Sir Robert Thompson, *Defeating Communist Insurgency* (Chatto and Windus, London, 1966); J.S. Pushtay, *Counter-Insurgency Warfare* (Free Press, New York, 1965); J.J. McCuen, *The Art of Counter-Revolutionary War* (Faber and Faber, London, 1966); Robert Taber, *The War of the Flea* (Paladin, London, 1970), p. 12.

11. Laqueur, *Guerrilla*, p. 375; Roger Trinquier, *Modern Warfare* (Pall Mall Press, London, 1964).

12. A. Short, *The Communist Insurrection in Malaya, 1948–60* (Frederick Muller, London, 1975), pp. 442–5; A. Short, 'The Malayan Emergency' in R. Haycock (ed.), *Regular Armies and Insurgency* (Croom Helm, London, 1979), pp. 53–68.

13. N.D. Valeriano and C.T.R. Bohannan, *Counter-Guerrilla Operations: The Philippine Experience* (Praeger, New York, 1962); David Galula, *Counter-Insurgency Warfare* (Praeger, New York, 1964).

14. D.S. Blaufarb, *The Counterinsurgency Era: US Doctrine and Performance* (Free Press, New York, 1977).

15. J.L.S. Girling, *People's War* (Allen and Unwin, London, 1969).

16. Cincinnatus (C.B. Currey), *Self-Destruction: The Disintegration and Decay of the US Army during the Vietnam War* (W.W. Norton, New York, 1981), p. 60.

17. Baylis, 'Revolutionary Warfare', pp. 132–51.

18. A. Horne, 'The French Army and the Algerian War, 1954–62' in Haycock, *Regular Armies*, pp. 69–83.

19. Frank Kitson, *Low Intensity Operations* (Faber and Faber, London, 1971).

20. T.L. Heyns, 'Will Afghanistan become the Soviet Union's Vietnam?', *Military Review*, October 1981, pp. 50–9.

21. Laqueur, *Guerrilla*, p. 165; A. Bennigsen, 'The Soviet Union and Muslim Guerrilla Wars, 1920–81: Lessons for Afghanistan', *Conflict* 4/1983, pp. 301–24.

22. D.C. Isby, 'Afghanistan, 1982: The War Continues', *International Defence Review* 11/1982, pp. 1523–8; J. Collins, 'Soviet Military Performance in

Afghanistan: A Preliminary Assessment', *Comparative Strategy* 2/1983, pp. 147–68.

23. R. Marston, 'Resettlement as a Counter-Revolutionary Technique', *Journal of the Royal United Services Institute* 124/4, 1979, pp. 46-9.

24. Collins, 'Soviet Military Performance', pp. 147–68.

1 THE BRITISH ARMY: THE DHOFAR CAMPAIGN, 1970–1975

John Pimlott

Britain's armed forces enjoy an enviable reputation for success in the notoriously difficult art of counter-insurgency. Since the end of the Second World War in 1945 there has hardly been a time when such forces have not been involved in at least one (and quite often more than one) campaign against armed insurgents intent upon the overthrow of existing political structures within a state under British rule or protection. The fact that none of these campaigns may be termed a failure for the Security Forces — indeed, many may be described as remarkable successes — makes a study of British techniques essential to any assessment of counter-insurgency in the modern age, particularly when it is borne in mind that other Western powers have spent the same period facing humiliation and defeat at the hands of similar insurgent groups.

The reasons for this degree of British success are many and varied. Among the most important must be the wealth of experience already amassed in what Major-General C.E. Callwell called 'small wars'[1] — those campaigns of colonial policing in which the British Army and, after 1920, the Royal Air Force as well, were constantly involved. The concept of politico-military insurgency may be relatively new, but many of the techniques involved, particularly of guerrilla warfare, are merely adaptations of traditional rebel tactics against which the British had often fought, and it is surprising how many of the methods used in what is now known as counter-insurgency have their origins in such an imperial past. During the guerrilla phase of the Boer War between 1900 and 1902, for example, the principle of cutting the insurgent off from his sources of support and supply was firmly established, as British forces responded to hit-and-run 'commando' raids by isolating the Boer homeland behind elaborate 'blockhouse' defences before systematically destroying the 'safe bases' of farms and homesteads, burning crops and removing the families of enemy fighters. At the same time, captured commandos were persuaded to work for the British as guides or trackers, special mounted infantry units were raised to play the Boers at their own game and offers of reward or amnesty were used to weaken guerrilla resolve. Such techniques — some of which may be traced back as far as the American War of Independence (1775–83) and the campaigns in India throughout the nineteenth century — were part of a 'normal' British response to colonial revolts which was readily available for use in 1945.

Nor was the experience confined just to the policing role, for by the end of the Second World War significant numbers of British soldiers and colonial policemen had equal familiarity with the actual conduct of guerrilla warfare. Between 1916 and 1918 Colonel T.E. Lawrence had led the way with his work among the tribesmen of the Arabian peninsula, exploiting their knowledge of the ground and natural fighting skills to harass and wear down the Turks; in the late 1930s Orde Wingate, influenced in part by Lawrence's book *Seven Pillars of Wisdom*,[2] did much the same with Jewish settlers in Palestine, helping to organise 'Special Night Squads' of fighters to pre-empt Arab attacks on *kibbutzim*. During the Second World War, Special Operations Executive (the leaders of which had made a detailed study of IRA techniques in the Irish 'troubles' of 1916–21)[3] trained guerrilla groups for operations against the enemy in the occupied areas of Europe and the Far East, and British soldiers gained firsthand knowledge of the problems involved by serving with or leading such groups. In Malaya, for example, men of Force 136 helped to run the Malayan People's Anti-Japanese Army between 1943 and 1945: when that later became the Malayan Races' Liberation Army and conducted a campaign against continued British colonial rule, beginning in 1948, some of those men were still available to lend weight to the counter-insurgency operations. Others, having spent the war in such exotic units as the Special Air Service, Long Range Desert Group, Popski's Private Army or Wingate's Chindits, could add their knowledge of deep-penetration raiding and unconventional fighting to produce a body of skills of inestimable value against insurgency groups.

But there was more to success than mere experience for, as the French were to show in both Indochina (1946–54) and Algeria (1954–62), a dependence upon previous techniques can lead to the adoption of inapplicable responses. The British advantage lay

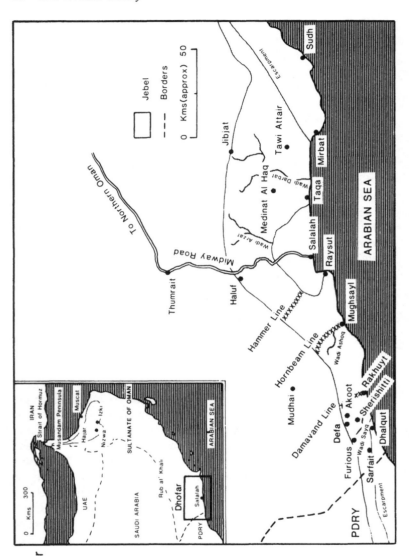

Map 1.1:
Southern Dhofar

in a tradition of flexibility, based upon the fact that throughout the colonial policing campaigns of the past they had been forced to make do with only limited resources. Global responsibilities had spread a relatively small volunteer army thinly on the ground and precluded the maintenance of a strategic reserve, while financial parsimony had made the soldiers aware of a need to husband their supplies of ammunition and equipment. Once presented with a revolt, therefore, the British were more likely to take a 'low profile' response, using their forces sparingly and searching for solutions which did not necessitate large expenditure of men or materiel; an approach which often made full use of local resources and involved close co-operation with existing civil authorities. At the same time, the wide range of threats to imperial rule — from the sophisticated armies of India to the stone-age tribesmen of parts of West Africa — and the different geographical conditions encountered, produced a constant need to adapt responses to fit local circumstances and avoided the development of a stereotyped 'theory' of policing.

Thus by 1945 the British already possessed the three important characteristics of experience, appropriate military skill and flexibility and, as they faced a plethora of threats to their rule or influence — in Palestine (1945–8), Malaya (1948–60), Kenya (1952–60), Cyprus (1955–9), Brunei and Borneo (1962–6), Radfan and Aden (1963–7), Northern Ireland (1956–62 and post-1969) and Dhofar (1970–5)[4] — these acted as a foundation upon which to build a distinctive pattern of counter-insurgency. What emerged was never a theory, elaborately compiled and rigidly adhered to in the manner of, say, the French *guerre révolutionnaire*, but a series of responses which, when adapted to fit specific conditions, proved successful in maintaining at least a measure of political stability, even under the pressure of strident nationalism or communist revolutionary warfare.

The key to this pattern was an appreciation of the fundamental point that, regardless of the nature of the military threat, all insurgencies have one thing in common — a desire on the part of the insurgents to take over political power. This will be sought not through direct military confrontation — the insurgents will lack the strength, equipment and skills to destroy the Security Forces in immediate open battle — but by a process of gradual subversion, persuading (or forcing) the people of a state to support the aims of the insurgency or, at least, to stop supporting the government. Either way, the centre of state power will be progressively isolated, and although the Security Forces protecting that centre will have to be worn down using guerrilla techniques, military actions by the insurgents will always remain subordinate to the overriding aim of political usurpation.

Once this has been recognised by the government, the response should reflect the pattern, with political actions, designed to prevent the insurgents gaining the support they need, taking priority over military moves to counter the guerrillas. After all, if the threat is political, then the long-term solution has also to be political: the role of the Security Forces should be to create an atmosphere in which guerrilla attacks do not disrupt the process of legitimate political rule. The extent to which the British have absorbed these basic points may be seen in the writings of Sir Robert Thompson, a soldier and administrator with wide experience of counter-insurgency. Of the five 'principles' that he lists as essential — that the government under threat should have a clear political aim, should function in accordance with the law, should establish an overall plan in which political, social, economic and military responses are carefully laid down, should give priority to defeating political subversion and should ensure that its own base areas are secure before mounting a military campaign[5] — four are essentially political in character and clearly dominate events.

But this is not something which emerges naturally, particularly if the Security Forces are not used to close political control or resent the imposition of constraints on their actions. One of the first moves in most British counter-insurgency campaigns has therefore been to impose civilian control using special legal powers provided to the politicians, often through the declaration of a State of Emergency. Control may be achieved by the simple expedient of combining the top civil and military posts in one man — in Malaya, for example, General Sir Gerald Templer acted as both High Commissioner and Director of Operations between February 1952 and June 1954 — or, more usually, by creating special committees containing appropriate representatives under civilian chairmanship. These committees are then extended downwards to regional and even village levels, ensuring political primacy throughout and co-ordinating civil and military responses at every stage.

Co-ordination is not much use, however, without detailed

knowledge of the threat and the committees are usually expected to place great emphasis upon the collection and collation of intelligence about the insurgents. At government level this invariably leads to the creation of a centralised intelligence department — that in Cyprus, for example, was established in 1955, even before a State of Emergency had been declared — although its success in building up an accurate overall picture of the insurgency will depend upon information provided by subordinate bodies. The most valuable of these should be the civil Police Force, using its knowledge of the country and contacts with the people to monitor events at grass-roots level, but quite often the fact that an insurgency has developed at all implies that the Police are weak. Some Western powers have responded by replacing them with military or para-military units, thereby losing the advantage of local knowledge; the British — perhaps in response to the general shortage of resources — prefer to rebuild the civil force through reform or the drafting in of more experienced personnel. In Northern Ireland, for example, the RUC was reorganised as a result of the Hunt Committee Report in 1969, while in Malaya 500 former members of the Palestine Police, already well-versed in counter-insurgency techniques, were recruited and placed under the command of Sir Arthur Young from the City of London force. The Army obviously has a part to play, providing tactical intelligence gleaned from cordon-and-search operations, patrols and the interrogation of captured enemy personnel, but the overall aim is to restore civilian policing as soon as possible, both to improve the flow of intelligence and to recreate an air of normality in the threatened state.

This should provide the government with a sound base from which to begin the process of countering subversive activity and the British priority has always been to isolate insurgents from their support among the people. The most obvious method is by straightforward propaganda, showing film-strips, distributing leaflets or, if possible, using the media to deliver the basic message that the *status quo* is worth supporting, although without more tangible proof of government concern for the people this is rarely sufficient to ensure success. All insurgencies have at their core a grievance, either real or contrived, which the insurgents exploit to gain support, and if this can be recognised and moves made by the government to satisfy it, the people will probably be tempted to shift their allegiance. In some cases radical change may be necessary — in both Malaya and Kenya a key element in the defeat of the insurgents' popular anti-colonial appeal was a British announcement that state independence would be granted according to democratic principles — but this often takes time to effect and may need to be allied to more immediate improvements. If, for example, poverty-stricken areas are provided with food, water supplies, medical aid and educational facilities, the population will soon realise that the government has far more to offer than insurgents who lack the resources to carry out their promised reforms. The aim should be to give the people a vested interest in the existing administration of the state: in Templer's words, to influence their 'hearts and minds'.

But the people must be assured that, having denied their support to the insurgents, they will be adequately protected against guerrilla retribution. The stationing of military or police detachments in the threatened areas is clearly one way of ensuring this, although in British campaigns a shortage of manpower and the need to maintain an effective mobile force for anti-guerrilla operations usually renders this impossible on a large scale. An alternative is to use local resources, raising militia or 'Home Guard' units from among the people, and despite an element of risk — after all, the government could be arming the very guerrillas it is trying to counter — this works if the people feel they are helping to protect something worthwhile. In Kenya, for example, the formation of 'Village Guard' detachments freed Security Forces from the static defence role and effectively protected tribal settlements against Mau Mau attack, even though the recruits were drawn from among the Kikuyu people in whose midst the insurgency lay. An added advantage of such local forces is that their creation implies government trust, and this may be crucial in ensuring support.

A more drastic form of protection is physically to remove the people from the areas of insurgent activity, resettling them in defended locations some distance away. It is a policy fraught with potential danger for the government, which may be accused of oppression, but if it is carried out humanely and the people are tempted by guarantees of security and personal gain, a notable degree of success may be achieved. By the end of 1951 in Malaya, for example, over 400,000 Chinese 'squatters', among whom the insurgents had organised support, were moved from squalid settlements on the jungle fringe into specially-constructed 'new villages', given land and provided with medical and educational facilities. Protected by militia units and wire defences, few squatters made the effort to re-establish contact with the insurgents they had left behind, and when this was coupled to a 'food denial' programme, whereby supplies provided to the new villages were carefully monitored and controlled, the insurgents soon found themselves isolated and alone.

Faced with such a situation, the insurgents will be forced onto the defensive, devoting their energies to personal survival rather than subversive activity, and it is at this point that the Security Forces begin to mount a sustained military campaign. The pattern was established by Lieutenant-General Sir Harold Briggs, appointed Director of Operations in Malaya in 1950, and he saw as the first priority the need to secure bases within the main centres of population, into which the insurgents might retreat, organising cordon-and-search operations in areas where resettlement was impracticable. Inevitably, innocent people will be affected by such moves — to prevent too much disruption the government may order the issue of identification (ID) cards which can be easily checked — but through a process of temporary detention and intelligence screening, suspects can be isolated and cells of 'passive support' for the insurgents eliminated. Operations of this description were carried out in Malaya, Kenya, Cyprus and Aden, while the destruction of Catholic 'no-go' areas in Belfast and Londonderry in July 1972 (Operation 'Motorman') had much the same effect.

This should complete the isolation of the insurgents, who will be forced to retreat into their own 'safe bases' deep in the countryside or among the slums of urban sprawl, and it is now the task of the Security Forces to pursue them, reducing the levels of guerrilla effectiveness and creating the atmosphere in which the government may initiate political solutions. If the 'safe bases' happen to be in remote or sparsely-populated regions, they may be designated 'Free Fire Zones' or 'Prohibited Areas' within which the Security Forces have free rein, but this is rare as the population has still to be protected and civilian casualties, however small, should be avoided to prevent unnecessary alienation. In such circumstances, the British tend to favour 'minimum force', although this may be part of a more general 'carrot and stick' approach, offering concessions to guerrillas or pockets of insurgent supporters who change sides, while threatening a destructive response if they persist in defying the government. In both Malaya and Cyprus, villages suspected of harbouring insurgents were collectively fined or subjected to stringent curfews; restrictions that could only be lifted if evidence of loyalty (usually in the form of intelligence) was forthcoming. On a more personal level, guerrillas may be tempted by offers of reward or amnesty and, as a symbol of government trust, may even be recruited into special units and sent back into remote or inaccessible areas to hunt down their erstwhile colleagues. In Kenya, Major (now General Sir Frank) Kitson achieved remarkable success in this respect using 'pseudo-gangs' of surrendered Mau Mau.[6]

But in the end it is the army and police who have to track down the remains of the guerrilla groups, and this often turns out to be the most gruelling aspect of the campaign. Again, a general shortage of manpower forces the British to deploy their units carefully, in ways that will achieve maximum effect, and this usually means taking on the enemy in his own environment, be it the jungles of Malaya or Borneo, the desert wastes of Radfan or the urban areas of Cyprus, Aden or Northern Ireland. This in turn shows the importance of flexibility, something made easier by the adaptability of the British soldier and the expertise of his training teams. Indeed, the high incidence of counter-insurgency since 1945 has produced a strong continuity of experience and skill throughout the British Army: by the time of the initial deployment to Northern Ireland in 1969, for example, it was not unknown for an infantry battalion to contain men (particularly senior NCOs) who had already fought in three or four different campaigns. This was a priceless advantage, for despite occasional evidence that each campaign began with a painful process of relearning the lessons of its predecessors, it created a repository from which to draw the strength to carry on.

The pattern of British counter-insurgency is therefore well established, based upon experience, skill and flexibility and containing the key elements of political primacy, insurgent isolation, intelligence and appropriate military response. The overall aim is to reverse the process of insurgency, with the government gradually assuming the advantages of surprise, tactical initiative and popular support which the insurgents initially enjoyed. In the end it should be the Security Forces who decide where and when to fight and it should be the government which sets the pace of political change. The pattern may not always produce complete success — in Cyprus a political compromise had to be accepted, in Aden economic pressures in Britain forced withdrawal before the campaign had fully developed, in Northern Ireland a political solution has still to be found — but it has enjoyed a better record than its counterparts in other Western states. Nowhere is this more apparent than in the British-inspired response to insurgency in the Omani province of Dhofar between 1970 and 1975.

The Sultanate of Oman is situated in the extreme southeast corner of the Arabian peninsula. Covering about 310,800 sq.km (120,000 sq. miles), it is bordered in the northwest by the United Arab Emirates — a strip of whose land actually divides the bulk of Oman from the sultan's outlying territory of the Musandam peninsula, overlooking the strategically-vital oil route through the Strait of Hormuz — in the west by Saudi Arabia and in the southwest by the People's Democratic Republic of Yemen (PDRY). Most of the one million population live on the northeastern Batinah plain, sheltered by the mountains of the Hajar and looking out over the Gulf of Oman. Behind the Hajar is a huge expanse of desert, merging into the Rub al'Khali or 'Empty Quarter' of Saudi Arabia, and although it is from this region that oil has been commercially exploited since the 1960s, communications are poor, with only a few rudimentary land routes between the Batinah plain and the western province of Dhofar, 800 km (500 miles) away.

Dhofar's isolation is further enhanced by its geography, which makes the province appear rather like an island, surrounded by the desert to north and east, the Arabian Sea to the south and extremely rough terrain to the PDRY border in the west. Within these boundaries, the topography is remarkably diverse. On the coast, centred upon the town of Salalah, is a crescent-shaped strip about 60 km (37 miles) long but never more than 14 km (9 miles) deep, which enjoys the benefits of the southwest monsoon (the Khareef) between June and September, producing a level of fertility and vegetation growth reminiscent of more tropical climes. Not surprisingly, the main settlements of Dhofar — Mirbat, Taqa, Salalah and Rakhyut — are situated on this coastal plain, although before the late 1970s they were rendered virtually inaccessible by the rolling surf of the Arabian Sea and the mountainous hinterland of the Jebel.

Indeed, it is the Jebel which dominates Dhofar, extending some 240 km (150 miles) from east to west, parallel to the coast, and rising sharply from the plain to heights of about 900 m (3000 ft) by means of extremely steep escarpments. Access to the mountains is marginally easier in the east, where the Jebel merges with the central desert of Oman, and they can be approached from the north across a gravel plain known as the Negd, but elsewhere the existence of deep, sheer-sided wadis, often hundreds of metres wide, hinders movement, particularly in the west, where the escarpment is more broken. Before the insurgency, the only land route across the Jebel was the Midway Road, a rough track leading from Thumrait on the Negd down to Salalah on the coast. During the monsoon, rain and mists affect much of the central and eastern Jebel, producing luxuriant plant growth; at other times the temperatures soar and the vegetation dies down. The indigenous population — about 10,000 mountain tribesmen collectively known as jebalis — exist according to the season, rearing cattle during the monsoon to see them through the drought, and their dependence upon water has tended to make them nomadic. They are fiercely independent and bear little similarity in language, dress or culture to the estimated 20,000 population of the Salalah plain. It was among the jebalis — ideal guerrilla fighters in classic guerrilla terrain — that the insurgency began in the early 1960s.

The reasons for their revolt are not difficult to understand, being part of a pattern of opposition to the rule of the sultan, Said bin Taimur. He had come to power on the abdication of his father in 1932, inheriting a backward, divided and debt-ridden country in which his authority was weak and his survival dependent upon British support. Determined to consolidate his rule, particularly at the expense of the religious leader, the Imam of Oman, who held sway among the tribes of the interior, Said had adopted a policy of deliberate parsimony, backed by draconian laws. By the early 1960s his position as head of state was certainly more secure — with British military aid he had defeated a revolt by the Imam and his followers in the Hajar (1957–9)[7] — and, with the promise of oil revenues in the near future, the country's finances were on a sounder footing, but Said had shown little regard for the well-being of his people. Oman was, to all intents and purposes, a feudal state, subject to the whim of the sultan and, as he distrusted all signs of progress or development, conditions were poor. No one was allowed to leave the country (nor, indeed, to move from one area to another within it) without Said's permission, and this was rarely granted; there were only three state-run schools in the whole of Oman, their exclusively-male pupils chosen by the sultan and denied education beyond a primary level; the only hospital was run, under sufferance, by an American Mission, yet diseases such as malaria, trachoma and glaucoma were endemic; and all symbols of the decadent twentieth century — from medical drugs and spectacles to books and radios — were banned. Said himself was a virtual recluse, living in his palace at Salalah, but despite his presence in Dhofar, the province suffered just as much as the rest of the state. Said had a particular dislike for the jebalis, describing them as 'cattle thieves' and fearing their traditional independence. As a result, he refused to sanction any policies which might benefit the mountain tribes.[8]

Many Omanis, desperate for education and money, left the country (illegally) and travelled to other Gulf states, joining local armed forces or finding work during the oil boom of the 1950s. Made aware of the advantages of twentieth-century life, their resentment at the government of their own state grew and, among the Dhofari exiles especially, active opposition began to develop, fuelled by the twin influences of Arab nationalism and 'scientific socialism'. In the early 1960s, having gained money and support from such 'front' organisations as the League of Dhofari Soldiers and the Dhofar Charitable Association, Sheikh Mussalim bin Nufl of the Al-Kathiri tribe founded the Dhofar Liberation Front (DLF), a small guerrilla group backed by the exiled Imam of Oman and the Saudi Arabians, dedicated to the overthrow of the sultan. As early as December 1962 sabotage was carried out at RAF Salalah and four months later oil-company vehicles were ambushed on the Midway Road. Bin Nufl then withdrew — he was sent by the Saudis to Iraq for more specialised guerrilla training — but he returned to the Jebel with about 50 followers in 1964 to resume his attacks on oil-company holdings.[9]

The sultan's response was weak, chiefly because he lacked both information about the threat and the means to gain that information. His army comprised only two British-officered battalions — the Northern Frontier Regiment (NFR) and the Muscat Regiment

(MR) — neither of which was stationed in Dhofar nor contained Dhofaris, and when the Commander, Sultan's Armed Forces (CSAF), Colonel Anthony Lewis, led an NFR company onto the Jebel in search of the DLF in October 1964, he faced enormous problems. Approaching the mountains from the north, Lewis had, in his own words, 'a cold start for getting to know the enemy, the inhabitants and the terrain'.[10] Fortunately the DLF was just as weak, lacking men, materiel and an organised infrastructure, and no contacts were made, but it was a poor start to a counter-insurgency campaign. Better results were achieved in May and June 1965, when two MR companies patrolled the wadis above Salalah in Operation 'Rainbow' — on two separate occasions DLF groups were found and defeated — but once the monsoon began, SAF troops were withdrawn and the mountains left to the rebels. By that time, after a congress held in early June, the DLF had become more structured, electing an 18-man executive and issuing a manifesto calling for the overthrow of the sultan and an end to foreign influence. These were appeals which found ready support among the jebali tribes.

During the 1965 monsoon the DLF consolidated defensive positions astride the Midway Road, aiming to prevent SAF reinforcement while they captured the settlements of Taqa, Mirbat and Mudhai. But they were still too weak to succeed, and although the sultan made no moves to counter growing rebel influence on the Jebel — on the contrary, in April 1966 he reacted to an attempt on his life by ordering a virtual blockade of the mountains, alienating hitherto-uncommitted jebalis who suddenly found themselves cut off from markets on the coast — the war quickly devolved into a stalemate. SAF patrols continued to probe the wadis, occasionally making contact with rebel groups, but no permanent bases were established in the mountains and military operations beyond the Salalah plain ceased altogether with the onset of the monsoon; the rebels, despite their new-found influence, lost their leader (bin Nufl was badly wounded during an attack on Mudhai in February 1966) and faced increasing problems of supply as Saudi Arabian support for the revolt waned and the supposedly-safe base at Hauf, over the border in what was still Aden Protectorate, was cleared by British troops. Neither side seemed able to gain the strength needed to achieve military victory and by 1967 Dhofar was strangely quiet. The revolt appeared to have died of natural causes, although in reality the DLF was using the lull to strengthen its political roots among the jebali people.

But the lull was only temporary, for in November 1967, as British troops withdrew from Aden, the nature of the revolt began to change. The new Marxist government of what was now the PDRY was swift to offer support to the Dhofari rebels, reopening Hauf as a safe base beyond the reach of the SAF and channelling supplies, mostly of Chinese or Soviet origin, onto the Jebel. Nor was this all, for with the supplies came a new breed of insurgents, heavily imbued with the ideology of the PDRY, fresh from an apparent victory over the British in Aden and intent upon a Marxist revolution throughout the Gulf. Their effect was immediate. At a second DLF congress, held in August 1968, the nationalist followers of bin Nufl were largely ousted from the 18-man executive and the movement renamed the Popular Front for the Liberation of the Occupied Arabian Gulf (PFLOAG — amended in 1971 to denote Popular Front for the Liberation of Oman and the Arabian Gulf). A new 25-man General Command was set up under Muhammad Ahmad al-Ghassani, dedicated to the spread of socialist revolution. In celebration, a fresh wave of attacks was mounted against the SAF: Salalah was hit by mortar fire (an incident which led to the construction of defended 'hedgehog' positions between the town and the escarpment in an effort to force the mortar teams out of range of the airfield) and Mirbat was attacked. With an estimated 2000 active fighters available, backed by up to 3000 jebali 'militia', the PFLOAG was a formidable force and one which outnumbered the SAF stationed in Dhofar.

Meanwhile the SAF had not been entirely idle, receiving something of a boost in April 1967 with the appointment of a new CSAF, Brigadier Corin Purdon. Together with the experienced commanding officer of the NFR, Colonel Michael Harvey, he began to revise the military strategy of the campaign, concentrating in 1968 on limited areas of Dhofar with the aim of bringing the rebels to battle and breaking up their forces into smaller, less dangerous groups. Some limited success was achieved in the east, but an attempt to interdict the supply route from Hauf as it entered western Dhofar failed and, in the continued absence of any political moves to tempt jebalis over to the sultan's side, no permanent advantages could be gained. Although Said was sufficiently alarmed to sanction a modest SAF modernisation programme, ordering helicopters, Strikemaster jets and semi-automatic rifles in an effort to counter the more sophisticated equipment of the PFLOAG, he still refused to offer any concessions to the mountain tribes, despite appeals to do so from his British advisers. As PFLOAG strength increased after August 1968, the SAF withdrew completely from western Dhofar, leaving rebel supply lines untouched and the people of the Jebel unprotected. Coastal towns came under renewed pressure, with both Rakhuyt in the west and Sudh in the east falling to the insurgents. The sultan's power by early 1970 was restricted to Taqa, Salalah and Mirbat, with only tenuous communications between the three positions. When it is added that in June 1970 a new organisation, the National Democratic Front for the Liberation of Oman and the Arabian Gulf (NDFLOAG), attacked the towns of Izki and Nizwa in northern Oman, the isolation and unpopularity of Said bin Taimur may be easily appreciated.

In the event the NDFLOAG was defeated, losing its most important leaders as one of its attacks went wrong, but this did nothing to disguise the desperate situation elsewhere in Oman. It was fully recognised in early 1970 by Lieutenant-Colonel John Watts, commanding officer of the British 22nd Special Air Service Regiment (22 SAS), as he conducted a preliminary reconnaissance of Dhofar in response to Said's rather belated request for allied aid. Watts was 'horrified' by what he found:

> The road was cut and the only resupply was by air or sometimes by sea. . . There were no Dhofaris in SAF, which was virtually an army of occupation. Everybody on the jebel was with the enemy, some convinced, some out of boredom, some intimidated: SAF had only a few jebali guides. It was crazy — we were on a hiding to nothing. . .[11]

All the basic ground rules of counter-insurgency had been ignored or broken: the sultan had made no attempt to understand the rebel grievances nor to counter them with offers of reform (as late as April 1970 Said's only comment to the new CSAF, Brigadier John Graham, was that the jebalis were 'evil and dangerous men — I want you to destroy them'[12]); there had been no government-sponsored propaganda designed to persuade jebalis to forsake the rebel cause; military operations, based upon scanty or useless intelligence, had been conducted in a vacuum, lacking an overall achievable aim or a political context; the SAF was weak and poorly equipped, leaving the rebels free to consolidate supply lines and mountain positions. In short the sultan was perilously close to defeat. Little wonder, therefore, that on 23 July 1970 his son, Qaboos bin Said, engineered his deposition in a near-bloodless palace coup, taking power in a desperate attempt to secure his inheritance. It proved to be a turning-point in the campaign.

Qaboos, Sandhurst-trained and supported wholeheartedly by the British, immediately provided the political leadership so desperately needed. Within 24 hours of the coup he had set up an Interim Advisory Council, chaired by the Defence Secretary and containing a number of British advisers, including the CSAF. Its first action was to invite Qaboos' uncle, Sayyid Tariq bin Taimur, to return from exile and become Prime Minister; when he arrived, he began to establish a modern central administration, setting up four new Ministries — Education, Health, Interior and Justice — in August 1970. Other exiled Omanis were tempted to return by offers of jobs in the new government, restrictions on movement within the state and to selected foreign countries were lifted, political prisoners released and, amid genuine displays of popular acclaim, a development plan announced, projecting the provision of schools, clinics, houses and roads for the whole of Oman. At the same time, Qaboos began to search for international recognition and aid, eventually joining both the United Nations and the Arab League in 1971, and, more specifically, turning to Britain for advice and assistance in the war against the PFLOAG.[13]

Politically, Britain was not in an ideal position to respond — since the withdrawal from Aden in 1967 both Labour and, more recently, Conservative governments had shown no desire to maintain involvement in the Gulf — but the growing importance of Oman as protector of the oil route through the Strait of Hormuz could not be ignored, especially as the PFLOAG made no move to disguise its communist ideology. Even so, there was never any question of a full-scale military commitment to Oman — economic pressures at home and the recent development of fresh troubles in Northern Ireland meant that large numbers of troops were not available, even if politically such a commitment had been acceptable — and from the beginning it was stressed that Britain would restrict its involvement to the provision of specialists and advisers. Thus, although units of the RAF Regiment, Royal Artillery, Royal Signals and Royal Engineers saw service in Dhofar after 1970, they were usually stationed in or around RAF Salalah, leaving direct participation in the counter-insurgency campaign on the Jebel to seconded or contract officers and NCOs attached to the SAF, a field surgical team (provided on a rotation basis by the British Army and RAF) and a BATT (British Army Training Team) from 22 SAS. Overall numbers were small — on average after 1970 about 150 seconded and 300 contract officers, backed by less than 50 medical and BATT personnel, were available at any one time — but they represented the means of applying to the Dhofar war all the lessons, techniques and experiences of British counter-insurgency campaigning so laboriously compiled since 1945. Their presence was crucial to eventual success.

This was apparent from the very beginning of the commitment, for once the political initiative of Qaboos' reforms had been announced and the first of the BATTs deployed, a general strategy for victory, based upon recommendations put forward by Lieutenant-Colonel Watts earlier in the year, was laid down. Watts had suggested an SAS campaign on five fronts, highlighting the need for intelligence collection and collation, an 'information service' to disseminate the government point of view to the jebalis, medical aid to the Dhofari people, veterinary facilities for the jebali cattle and a policy of directly involving Dhofaris in the fight for their province:[14] in other words, an attempt to apply the traditional British techniques of local knowledge, 'hearts and minds' and native participation to a war which had hitherto lacked all three. But, as Colonel Tony Jeapes, one of the first BATT commanders, has pointed out:

. . . these were stop-gap measures to plug holes until the Omani Government could provide its own people to do these tasks. The short term aim was to bring immediate relief to the people. The medium term aim was to train Omanis to take over these measures and then to hand over to them. The long term solution, however, was in the hands of the Omani Government, to better the lot of the Dhofari people by the development of resources and the construction of roads, wells, schools, clinics, mosques — everything that goes to make up a modern state. Military operations must simply be a means to that end.[15]

A familiar pattern of counter-insurgency soon began to emerge. On a political level, Qaboos clearly recognised the need for co-ordination and control, setting up a special Dhofar Development Committee (DDC) under the Wali (sultan's representative) in Salalah, to which high-ranking members of the SAF, police and interested government agencies belonged. Brigadier John Akehurst, commander of the SAF troops in Dhofar during the final months of the war, has described the workings of the DDC and assessed its value:

> The Committee met once a week, usually in the Wali's house, and each member in turn reported events of the past week and the intentions for the next. Any policy matters were then discussed and decisions promulgated at once . . . without fear of red tape. It made for successful management under wartime conditions at a time when Omani government departments were in their infancy and still unaccustomed to accepting responsibility for decisions.[16]

The DDC also ensured that all members of the counter-insurgent team retained awareness of the overall aim of the campaign and co-operated to achieve it. Akehurst defined his task in 1974 as 'to secure Dhofar for civil development,'[17] and although this has all the appearances of a military catchphrase, it does sum up the sultan's policy in a wider sense. Qaboos was aware from the beginning that the jebalis needed to be tempted rather than forced into supporting the *status quo*, and the best way to do this, as the British could bear witness, was to provide the people with material benefits which the insurgents could never hope to match — the 'roads, wells, schools, clinics, mosques' mentioned by Jeapes. These were to be organised by what were known as CATs (Civil Action Teams), usually comprising 'a leader, schoolmaster, medical orderly and shop-keeper'[18] who would spearhead a process of development and modernisation, principally on the Jebel. The idea was that a CAT would appear in a chosen tribal area and establish a 'centre', digging a well to provide constant supplies of water to the local people and their livestock, initiating rudimentary education and medical schemes and generally showing to the jebalis the advantages of life under the sultan. Hopefully, small permanent settlements would grow up around each centre, containing tribesmen who were loath to continue supporting the rebels for fear of losing their new government-sponsored benefits. The CAT would then supervise the building of roads to link the centre with the outside world and offer long-term aid through improvements to cattle stock and the provision of markets for jebali goods. Eventually each centre would become self-sufficient, the tribesmen would be well-motivated to remain loyal to the sultan and the insurgents would lose their popular appeal.

But in 1970 this could only be a dream, for although the idea was undoubtedly sound, the fact that the Jebel was in PFLOAG hands precluded the immediate deployment of CATs. The problem, at this stage, was a military one, centred upon the need to find and exploit weaknesses in the rebel camp which would allow the establishment of permanant SAF bases on the Jebel. Before this could even be contemplated, however, SAF commanders had to gather intelligence about the insurgents and their organisation. A breakthrough came, quite fortuitously, in September 1970, provided to a large extent by the PFLOAG itself. The shift to Marxism after 1968 had not been popular on the Jebel, chiefly because the new ideology undermined two of the fundamental aspects of jebali life — the Moslem religion and the tribal system. Young communists, fresh from training camps in the PDRY, Iraq and the Soviet Union, tried to destroy all vestiges of such religious and social cohesion, often by means of intimidation and violence, only to encounter entrenched opposition. When they moved to disarm dissident groups on the eastern Jebel on 12 September, a fire-fight ensued and 24 of the most experienced ex-DLF guerrillas, led by Salim Mubarak, chose to surrender to the sultan rather than bow to the PFLOAG. It was a windfall which the SAF could not afford to waste: instead of punishing these ex-rebels, government officers welcomed them and began a careful process of questioning. Mubarak and his men provided the first real insight into the rebel infrastructure.

There was more to the incident than that, however, for the realisation that many jebalis opposed the PFLOAG on religious grounds provided the SAF with a propaganda theme of great potential value. It was exploited immediately by the 'information service' experts of 22 SAS, who arrived in Dhofar just as

Mubarak surrendered.[19] Tasked to 'bring the truth to the people of Dhofar', they organised air-drops of leaflets onto the Jebel, the provision of illustrated 'fact sheets' for prominent display in towns and markets and, most importantly, the establishment of Radio Dhofar, broadcasting the government point of view throughout the province. Cheap transistor radios were made available to the people and a special slogan — 'Islam is Our Way, Freedom is Our Aim' — was composed and constantly used, stressing the religious aspect of the struggle. At first the problems were immense — few people on the government side, for example, were familiar with the jebali language — but over the succeeding few years steady progress was made. By 1975 Radio Dhofar was widely regarded as a source of reliable, accurate information and its pro-PFLOAG rival, Radio Aden, lay discredited. In the fight for Dhofari 'hearts and minds', this was an important asset.

Mubarak and his followers — officially designated SEPs (surrendered enemy personnel) — also provided another advantage: on the suggestion of the SAS, they were persuaded to join the sultan's fight against their erstwhile colleagues, forming the core of an anti-guerrilla unit known as Firqat Salahadin. Put together in late 1970, it was actively employed as early as 23 February 1971 when, with SAS help, it recaptured the eastern town of Sudh without firing a shot. Thereafter, as more guerrillas responded to the continued anti-Moslem/anti-tribal violence of the PFLOAG and surrendered, tempted by government offers of amnesty and promises of no punishment, the number of firqat units was increased. By late 1974 an estimated 1500 firqat fighters (about 80 per cent of them SEPs) were available in 18 separate groups,[20] although it would be wrong to imagine that their organisation was straightforward. Initially, the intention had been to recruit on a multi-tribal basis (Qaboos, like the rebels, had reason to fear the power and independence of the tribal system), but, after mutinies and intra-firqat squabbles, it was decided to raise each group according to tribal affiliations. This proved to be an astute move, ensuring cohesion, strong leadership and local knowledge, and when the process was taken one step further, making each firqat responsible for the security of its own tribal area on the Jebel, the motivation to defeat the PFLOAG and liberate kinfolk was dramatically increased. Although SAS-led firqat continued to take part in more general military operations, their evolution as a territorial militia or 'Home Guard' proved to be much more valuable in the long term, driving a wedge between the insurgents and the people in the mountain areas and showing, for the first time, that jebalis were an integral part of the government campaign.[21]

But these were all aspects of the counter-insurgency pattern which, although undoubtedly crucial to eventual success, could not begin to be effective until the SAF had re-established control on the Jebel. Qaboos realised the weakness of his forces as soon as he came to power and immediately authorised a policy of reorganisation and expansion. Within months a separate command — Dhofar Brigade — had been established under a British brigadier, recruitment to the SAF had been stepped up and new equipment (including that ordered by Said before the coup) had been issued. By 1974 Dhofar Brigade was an efficient and well-equipped organisation, comprising about 10,000 fighting men drawn from a wide variety of sources.

Of the four Omani battalions of the SAF — the NFR, the MR and the newly-raised Desert Regiment (DR) and Jebel Regiment (JR) — two were always stationed in Dhofar on nine-month rotational tours, joining two other battalions — the Frontier Force (FF) and Kateebat Janoobiya (KJ or Southern Regiment), both recruited from the ex-Omani territorial possession of Baluchistan, now part of Pakistan — which made up the permanent garrison. They were supported on the ground by elements of the Oman Artillery, deploying a mixture of 5.5 in medium howitzers, 25-pounder gun-howitzers, 75 mm light guns and 4.2 in mortars, by a detachment of the SAF Armoured Car Squadron, equipped with Saladins, and by the para-military Oman Gendarmerie. In the air, the Sultan of Oman's Air Force (SOAF) flew Strikemaster ground-attack jets, Skyvan short take-off and landing transports, light helicopters and liaison aircraft, while at sea any vessel of the Sultan of Oman's Navy (SON) which sailed into Dhofari waters automatically came under command. All of these units were officered by British seconded or contract personnel, the latter recruited by Airwork Services Limited, which also provided technicians to look after the more sophisticated equipment. BATT and firqat soldiers were also available, together with troops from Middle Eastern states which had responded to Qaboos' request for aid. In December 1973 an Imperial Iranian Battle Group (IIBG) began to arrive in Dhofar, to be stationed at Thumrait on the Midway Road, and, together with Jordanian engineers and Special Forces, they participated in many of the later operations of the war. Problems were not unknown — the constant roulement of Omani battalions, for example, disrupted the continuity of experience, while co-operation between the SAF and the Iranians was not always good — but compared to the situation under Said before 1970, Dhofar Brigade was a formidable force.[22]

An offensive against the PFLOAG began in early 1971, when the newly-rotated NFR established a base at Haluf, 32 km (20 miles) to the north of Salalah, in Operation 'Hornet'.[23] The intention was to break down a large enemy group known to be operating in the central area of the Jebel, and some success was achieved when elements of that group rather foolishly decided to stand and fight, suffering the attentions of both the SOAF and the Oman Artillery. By March they were no longer a viable force, enabling the SAF to mount further operations in the west, setting up a base at Akoot, close to a rebel supply dump in the Sherishitti caves. The capture of Sudh by SAS/firqat troops in February had already threatened PFLOAG control in the east, so that by the middle of 1971 Qaboos' forces had made their presence felt at points throughout the Jebel. But they still lacked the means to maintain permanent mountain bases: as soon as the monsoon began in June, all SAF units were withdrawn to the Salalah plain, where supplies and communications were reasonably secure, leaving the PFLOAG to regain their control.

This was a serious flaw in the government campaign, undermining all attempts to persuade jebalis to support the sultan. An anonymous tribesman summed up the problem when greeting an SAS/firqat patrol in late 1971:

> You say you will be here a long time. But what is a long time, one week, two weeks? And the *geysh* (government forces) have never stayed here during the monsoon. The Communists are here the whole time. As soon as you leave, they will come back and punish anyone who helped you.[24]

In such circumstances, the establishment of permanent bases on the Jebel was clearly essential, both to disrupt the rebel build-up and to allow the deployment of CATs. During the 1971–2 campaigning season a series of co-ordinated SAF operations took place with this aim firmly in mind.

The process began on 2 October 1971 when, in Operation 'Jaguar', an SAS/firqat group moved onto the eastern Jebel to the north of Mirbat, taking Jibjat and pushing down both sides of the Wadi Darbat. Despite tough opposition, a permanent position was established at 'White City' (Medinat Al Haq), enabling the first of the CATs to set up a school, clinic and shop. When the government forces did not leave with the onset of the monsoon, local tribesmen began to believe the promises of civil development, spreading the word throughout the eastern Jebel. PFLOAG strength was further affected by a simultaneous offensive — Operation 'Leopard' — in the mountains close to Mughsayl in the west, for this disrupted the flow of rebel supplies from the PDRY into the eastern region. Once this had occurred, a third attack — Operation 'Panther' — exploited rebel weakness by moving out of Jibjat towards Tawi Attair, and although the new positions had to be abandoned in May 1972, significant damage had been inflicted on the PFLOAG infrastructure in the east. It was an encouraging start.

But not all SAF offensives enjoyed such success. On 15 April 1972 Operation 'Simba' was mounted in the far west, with DR soldiers being airlifted onto escarpment positions at Safait, close to the PDRY border. On the map, this was a logical move — supply routes across the border were forced by the terrain to hug the coast and SAF positions on the heights above would disrupt or even halt the flow — but in reality the operation was over-ambitious. Although the DR troops secured positions on top of the escarpment, they found that they could not see the coastal track and, when they tried to move down to a lower plateau, the roughness of the terrain and lack of water forced them back. They immediately came under artillery fire from within the PDRY and a siege began. The SAF had the advantage of air power, enabling the Sarfait defenders to be relieved and resupplied, but the value of the position was questionable, particularly as it tied down badly-needed forces without affecting the rebel supply line.

Even so, by the beginning of the 1972 monsoon, the military situation on the Jebel had been transformed. PFLOAG forces in the east had been seriously disrupted, the supply line north of Mughsayl had been interdicted and, although the Sarfait position had been contained, the ease with which the SAF had moved to secure it, using air power, augured ill for the future of the rebels. Realising the threat to their control, the PFLOAG leadership ordered a counterattack, choosing to mount a massed frontal assault on the eastern town of Mirbat, defended by about 30 men only, including an eight-man BATT under Captain Mike Kealy of 22 SAS. At dawn on 19 July 1972 over 200 PFLOAG fighters attacked behind a barrage of artillery and mortar fire, and it took impressive leadership from Kealy (who was subsequently awarded the DSO for his exploits), the commitment of SAS reinforcements and the impact of SOAF Strikemaster support to prevent disaster.[25] As it turned out, the battle was a major blow to the PFLOAG, forcing them to recognise their weakness in conventional operations against Dhofar Brigade — it was, in fact, the last time the insurgents attempted open warfare — but it had been a closely-fought contest.

The initiative now lay with the government, although it would appear that SAF commanders were slow to realise this. Qaboos laid down a list of military priorities for 1972–3 — to defend the Salalah plain, protect the border with the PDRY and open up the Jebel to more extensive civil development — but it was unadventurous and rather vague. This was reflected in the new campaigning season, for although operations continued in the east, with DR troops moving out of White City towards the Midway Road and elements of both the NFR and JR consolidating positions astride the 'Leopard Line' north of Mughsayl, the attacks achieved little of lasting value and the PFLOAG was allowed to recover. In March 1973 the rebels showed their strength by hitting RAF Salalah with rockets, damaging three helicopters and two Strikemasters and forcing the defenders to man new positions ('Dianas') closer to the escarpment. It was a sobering reminder that the PFLOAG was still an effective guerrilla force.

Fortunately the new CSAF, Major-General Timothy Creasey, recognised the need for a more positive strategy and in late 1973 he ordered Dhofar Brigade, commanded by Brigadier Jack Fletcher, to concentrate on two specific objectives — the destruction of rebel forces in the east and the permanent interdiction of the supply route from the west. These tasks were made easier in December 1973 when an Iranian parachute battalion — spearhead of the IIBG — reopened the Midway Road, establishing a series of defended positions ('Jasmines') which denied the area to the PFLOAG and drove a wedge through the centre of the Jebel. SAF units followed this up by constructing an 80 km (50 mile) wire and mine barrier— the Hornbeam Line — into the mountains to the north of Mughsayl in early 1974. This isolated the eastern Jebel even more and, as MR troops with firqat assistance began to clear the area to the east of the Midway Road, the rebels held another congress (August 1974). The fact that they decided to change their name to the Popular Front for the Liberation of Oman (PFLO) implied that the former aim of sparking a revolt throughout the Gulf had been forced into second place behind that of sheer survival in Dhofar. It was a measure of SAF success.

Thus when Akehurst took over Dhofar Brigade in September 1974, he inherited a healthy military situation. At first he concentrated on clearing operations east of the Hornbeam Line — in October the FF reopened the coast road from Taqa to Mirbat, while other SAF units patrolled the area between Hornbeam and the Midway Road, setting up an intermediate barrier known as the Hammer Line — although this was only a preliminary to the far more difficult task of destroying rebel positions in the west. In December 1974 a new offensive was mounted, using the IIBG to conduct a two-pronged attack towards Sherishitti and Rakhuyt. It was not a complete success. Despite the reoccupation of Rakhuyt and the construction of another defensive barrier — the Damavand Line — the advance towards Sherishitti encountered stiff opposition and ground to a halt. A second offensive — Operation 'Dharab' — was hastily organised, with JR, firqat and SAS troops threatening Sherishitti from the north in an attempt to relieve the pressure on the Iranians, but this too experienced problems. When it began on 4 January 1975 the advance was slow, surprise was quickly lost and the enemy reacted with unexpected force. SAF morale recovered on 21 February when, in Operation 'Himaar', a rebel HQ was attacked and destroyed in the Wadi Ashoq, west of the Hornbeam Line, but it was apparent that more sustained and carefully-prepared offensives would be needed to defeat the substantial enemy force on the western Jebel. The appearance in Dhofar of regular troops from the PDRY and the continuing artillery fire from across the border, merely made the problem worse.

Creasey was replaced as CSAF in March 1975 by Major-

General Kenneth Perkins,[26] who quickly devised a plan for a final government offensive, to be mounted on 21 October, at the end of Ramadan. Code-named 'Hadaf', it envisaged setting up one more defensive barrier, anchored on the coastal town of Dhalqut, as far to the west as it was possible to go without coming under PDRY artillery fire. SAF troops would approach the heights above the Wadi Sayq, northeast of Sarfait, from positions at Defa and 'Furious' in the north, after which a heliborne assault would secure the Darra Ridge overlooking Dhalqut. To divert PFLO attention, subsidiary operations would be mounted from Sarfait towards the coast and from bases in the north and south towards Sherishitti. As a preliminary, the route to Defa was cleared in August.

Changes were forced on the SAF commanders in mid-October, however, for when the first diversionary attack began on the 15th, patrols of the MR succeeded, against all the odds of terrain, weather and mines, in advancing onto a lower plateau, nearer the coast. From here the rebel supply route could be clearly seen and, in an impressive display of flexibility, Akehurst threw 'seven months of planning and 40 pages of operation orders out of the metaphorical window'[27] and, backed by Perkins, ordered the push to continue down to the sea. Two SAF companies were moved into Sarfait as reinforcements and Qaboos authorised the SOAF, flying Hawker Hunter jets recently acquired from Jordan, to mount selected air strikes against artillery positions in the PDRY. The gamble paid off: PDRY troops hastily withdrew from Dhofar as their artillery support was silenced, leaving the PFLO isolated and alone. Thus when the second diversionary attack — Operation 'Saeed' — began on 17 October, with an Iranian force moving out of Rakhuyt to threaten Sherishitti from the south, opposition was slight. Four days later, Operation 'Hadaf' began on schedule and FF troops cleared the heights to the north of Sherishitti, effectively boxing-in the remnants of the PFLO. The caves were finally captured in November, by which time the advance onto the Darra Ridge had begun. By 4 December the PFLO was broken and retreating towards the PDRY, allowing Akehurst to send a signal of victory to Qaboos: 'I have the honour to inform your Majesty that Dhofar is now secure for civil development.'[28]

Clearing operations continued throughout the Jebel until mid-1976, but there was no denying that the SAF, with British,

Iranian and Jordanian help, had achieved a notable victory over the Dhofari rebels. This had been made possible through a combination of techniques — the imposition of centralised politico-military control, the evolution of an achievable aim, good intelligence, effective propaganda, the isolation of the insurgents from their sources of support and supply and the use of appropriate military skills — which were firmly based upon the traditional pattern of British counter-insurgency. But it would be wrong to leave the story there. What had been achieved was a *military* victory and, as the British have always stressed, that on its own does not guarantee lasting security. Akehurst had presented the sultan with a situation of military calm; it was up to Qaboos to exploit this to the full by introducing political, economic and social reforms which would ensure the loyalty of his people. Then, and only then, would the campaign be complete.

Subsequent events suggest that this has occurred in Oman.[29] The administrative reforms of 1970 have been extended to produce the infrastructure of a modern state, oil revenues have been devoted to modernisation and civil development and the people have begun to enjoy the advantages of well-organised medical and educational facilities. Communications have been improved, new industries have been introduced and a process of 'Omanisation', replacing foreign soldiers, managers and technicians with trained Omanis, has been initiated. In Dhofar itself, new towns have been built, connected by tarmac roads, while a graded track links the province to the rest of the state. A new port has been constructed at Raysut, 21 km (13 miles) to the west of Salalah; a power grid has been set up, enabling telephone and television facilities to be introduced; schools, a hospital and a host of clinics and dispensaries have been built. More importantly, the CAT scheme has provided jebalis with water, food supplies, medicines and education and their cattle-based economy has been vastly improved by cross-breeding and veterinary care. Indeed, beef is now a major export of Dhofar, giving the jebalis a strong vested interest in preserving the *status quo*. All of this has effectively reduced the chances of renewed rebel activity; something which has only been achieved by a co-ordination of military and political policies — the military to quieten the violence, the political to ensure that it does not re-emerge — which, above all else, forms the central core of the

British counter-insurgency pattern. Its success caused Jeapes to describe the Dhofar war as 'a model campaign'[30]: it is difficult to fault his judgement.

Notes

1. C.E. Callwell, *Small Wars. Their Principles and Practice* (HMSO, London, 1896).
2. T.E. Lawrence, *Seven Pillars of Wisdom* (Jonathan Cape, London, 1935).
3. See M.R.D. Foot, 'The IRA and the Origins of SOE' in M.R.D. Foot (ed.), *War and Society: Historical Essays in Honour and Memory of J.R. Western 1928–1971* (Paul Elek, London, 1973), pp. 57-69.
4. For easily-accessible information on these campaigns, see J. Paget, *Counter-Insurgency Campaigning* (Faber and Faber, London, 1967); G. Blaxland, *The Regiments Depart: The British Army, 1945–1970* (William Kimber, London, 1971), and the partwork *War in Peace* (Orbis, London, 1983–4).
5. The 'five principles' are taken from R. Thompson, *Countering Communist Insurgency* (Chatto and Windus, London, 1966), pp. 50–7.
6. F. Kitson, *Gangs and Countergangs* (Barrie and Rockliff, London, 1960).
7. For coverage of the 1957-9 campaign, see D. Smiley and P. Kemp, *Arabian Assignment* (Leo Cooper, London, 1975) and S. Monick, 'Victory in Hades: The Forgotten Wars of the Oman, 1957–1959 and 1970–1976, Part I', *Militaria. The Official Professional Journal of the SADF* (Pretoria), 12/3, 1982.
8. Said's regime is well covered in F.A. Clements, *Oman. The Reborn Land* (Longman, London, 1980), pp. 49–65.
9. See D.L. Price, *Oman: Insurgency and Development* (Conflict Studies, London, Paper no. 53, January 1975).
10. Quoted in Lieutenant-Colonel J. McKeown, RE, 'Britain and Oman: The Dhofar War and its Significance' (Unpublished M.Phil dissertation, University of Cambridge, 1981), p. 22.
11. Ibid., p. 46.
12. Ibid., p. 47.
13. See Clements, *The Reborn Land*, pp. 65–73.
14. Watts' 'five front' plan is covered in Colonel Tony Jeapes, *SAS: Operation Oman* (William Kimber, London, 1980), p. 31, and Tony Geraghty, *Who Dares Wins. The Story of the SAS, 1950-1980* (Arms and Armour Press, London, 1980), pp. 156–7.
15. Jeapes, *Operation Oman*, p. 31.
16. J. Akehurst, *We Won a War: The Campaign in Oman, 1965–1975* (Michael Russell, Salisbury, 1982), p. 54.
17. Ibid., p. 65.
18. This description of a typical CAT is taken from McKeown, 'Britain and Oman', p. 81. More detail on civil development in Dhofar may be found in Clements, *The Reborn Land*, pp. 91–108.
19. For details of 'information services', see Jeapes, *Operation Oman*, pp. 34–6.
20. These figures are taken from Price, *Insurgency and Development*, p. 11, and apply to late 1974. McKeown cites 'over 3,000' by the end of the war in December 1975, 'Britain and Oman', p. 56.
21. Details of firqat operations may be found in Geraghty, *Who Dares Wins*, pp. 172–7, and in Jeapes, *Operation Oman*, passim.

22. The description of Dhofar Brigade in 1974 is based on Akehurst, *We Won a War*, pp. 31–45.

23. The best general coverage of operations 1970–6 may be found in McKeown, 'Britain and Oman', pp. 50–95, Jeapes, *Operation Oman, passim*, and Akehurst, *We Won a War, passim*.

24. Jeapes, *Operation Oman*, p. 94.

25. The Battle of Mirbat is covered in detail in Geraghty, *Who Dares Wins*, pp. 161–72, and Jeapes, *Operation Oman*, pp. 143–58. Kealy died of exposure during an SAS exercise in Wales in February 1979.

26. Perkins has written his own very useful account of the last campaigns of the war: 'Oman 1975: The Year of Decision', *Journal of the Royal United Services Institute* (London), 124/1, March 1979, pp. 38–45.

27. Quoted in McKeown, 'Britain and Oman', p. 92.

28. Akehurst, *We Won a War*, p. 173.

29. Post-1975 developments in Oman are extensively covered in Clements, *The Reborn Land*, pp. 65–152.

30. Jeapes, *Operation Oman*, p. 11.

References

A. *General Works*

Barber, N. *The War of the Running Dogs. How Malaya Defeated the Communist Guerrillas, 1948–1960* (William Collins, London, 1971)

Blaxland, G. *The Regiments Depart. The British Army, 1945–1970* (William Kimber, London, 1971)

Callwell, C.E. *Small Wars. Their Principles and Practice* (HMSO, London, 1896)

Clayton, A. *Counter-Insurgency in Kenya. A Study of military operations against Mau Mau* (Transafrica Publishers, Nairobi, 1976)

Evelegh, R. *Peace-Keeping in a Democratic Society. The Lessons of Northern Ireland* (C. Hurst, London, 1978)

James, H.D. and Sheil-Small, D. *The Undeclared War. The Story of the Indonesian Confrontation, 1962–1966* (Leo Cooper, London, 1971)

Kitson, F. *Gangs and Countergangs* (Barrie and Rockliff, London, 1960)

Kitson, F. *Low Intensity Operations. Subversion Insurgency and Peacekeeping* (Faber and Faber, London, 1971)

Kitson, F. *Bunch of Five* (Faber and Faber, London, 1977)

Lawrence, T.E. *Seven Pillars of Wisdom* (Jonathan Cape, London, 1935)

Majdalany, F. *State of Emergency. The Full Story of Mau Mau* (Longman, London, 1962)

McCuen, J. *The Art of Counter Revolutionary War* (Faber and Faber, London, 1966)

Paget, J. *Counter-Insurgency Campaigning* (Faber and Faber, London, 1967)

Paget, J. *Last Post: Aden, 1964–1967* (Faber and Faber, London, 1969)

Thompson, R. *Defeating Communist Insurgency* (Chatto and Windus, London, 1966)

Townshend, C. *The British Campaign in Ireland, 1919–1921. The Development of Political and Military Policies* (Oxford University Press, Oxford, 1975)

War in Peace partwork (Orbis, London, 1983–4)

B. *Oman and the Dhofar War*

Akehurst, J. *We Won a War. The Campaign in Oman, 1965–1975* (Michael Russell, Salisbury, 1982)

Clements, F.A. *Oman. The Reborn Land* (Longman, London, 1980)

Fiennes, R. *Where Soldiers Fear to Tread* (Hodder and Stoughton, London, 1975)

Geraghty, T. *Who Dares Wins. The Story of the SAS, 1950–1980* (Arms and Armour Press, London, 1980)

Halliday, F. *Arabia Without Sultans* (Penguin, London, 1974)

Halliday, F. *Armed Struggle in Arabia: Counter Insurgency in Oman* (The Gulf Committee, 1976)

Jeapes, T. *SAS: Operation Oman* (William Kimber, London, 1980)

McKeown, J. 'Britain and Oman: The Dhofar War and its Significance' (Unpublished M.Phil dissertation, University of Cambridge, 1981)

Monick, S. 'Victory in Hades: The Forgotten Wars of the Oman, 1957–1959 and 1970–1976, Part I', *Militaria. The Official Professional Journal of the SADF* (Pretoria), 12/3, 1982

Monick, S. 'Victory in Hades: The Forgotten Wars of the Oman, 1957–1959 and 1970–1976, Part 2: The Dhofar Campaign, 1970–1976. Section A', *Militaria. The Official Professional Journal of the SADF* (Pretoria), 12/4, 1982

Monick, S. 'Victory in Hades: The Forgotten Wars of the Oman, 1957–1959 and 1970–1976, Part II (*sic*); The Dhofar Campaign, 1970–1976. Section B', *Militaria. The Official Professional Journal of the SADF* (Pretoria), 13/1, 1983

Perkins, K. 'Counter-Insurgency and Internal Security', *British Army Review* (London), no. 69, December 1981

Perkins, K. 'Oman 1975: The Year of Decision', *Journal of the Royal United Services Institute* (London), 124/1, March 1979

Price, D.L. *Oman: Insurgency and Development* (Conflict Studies, London, Paper no. 53, January 1975)

Purdon, C.W.B. 'The Sultan's Armed Forces', *British Army Review* (London), no. 32, August 1969

Smiley, D. and Kemp, P. *Arabian Assignment* (Leo Cooper, London, 1975)

Townsend, J. *Oman: The Making of the Modern State* (Croom Helm, London, 1977)

Tremayne, P. 'Guevara Through the Looking Glass, A View of the Dhofar War', *Journal of the Royal United Services Institute* (London), 119/3, September 1974

Tremayne, P. 'End of a Ten Years' War', *Journal of the Royal United Services Institute* (London), 122/1, March 1977

2 THE FRENCH ARMY: FROM INDOCHINA TO CHAD, 1946–1984

John Pimlott*

The French have not enjoyed a great deal of success in counter-insurgency since 1945. They have experienced defeat at the hands of a relatively unsophisticated peasant army in Indochina (1946–54) and faced the trauma of virtual civil war over insurgency in Algeria (1954–62). In the process they evolved and adopted a rigid 'theory' of response — known as *guerre révolutionnaire* — which not only proved to be unsuccessful in practice but also led to deep rifts in French society as the Algerian War dragged on, and even provoked attempts by the Army to interfere in the internal politics of the state on two occasions (1958 and 1961). Since 1962, there has been an understandable tendency not to become involved too deeply in counter-insurgency situations, leading to an apparent lack of long-term commitment in places such as Chad, where French troops have been deployed on no less than three separate occasions to little permanent effect. The record is therefore hardly an impressive one, although this should not make it unworthy of study. The French problems in Indochina and Algeria may have been unique and impossible to solve using methods adopted by other Western powers elsewhere, but the evolution and failure of a highly-structured approach, as epitomised by *guerre révolutionnaire*, deserves examination, if only to show the pitfalls and dangers involved.

In 1945 there was little obvious indication that such defeats would occur. Despite the failure of French armed forces at home in May and June 1940 and the subsequent humiliation of enemy occupation for four years, the colonial areas remained remarkably unaffected, occasionally being fought over but rarely causing trouble in their own right. Indeed, the relative ease with which the French reasserted their colonial authority at the end of the

* The author gratefully acknowledges the help provided by Lt.-Col. P.E.X. Turnbull in the preparation of this chapter.

Second World War suggested that little had changed and that the pre-war pattern of life would continue. In the 1930s, isolated voices claiming to represent movements for 'freedom from foreign domination' had been heard, but their influence had remained, on the surface, insignificant. Though such tendencies were encouraged during the period of enemy occupation or control, particularly in Indochina, the return to the *status quo* of 1939 did not offer any evidence that potential dissidents had gained in popularity or possessed the means to pose any threat — let alone of a military nature — to the colonial power. The only sign of trouble came in Algeria in May 1945, where celebrations to mark the end of the war in Europe flared into violence between Moslems and French settlers, but a swift and ruthless military reaction, which included the use of summary executions, air strikes and even naval bombardment of Moslem centres, soon restored order in favour of the French.[1]

Algeria was to remain quiet for another nine years, reinforcing the view that traditional methods of colonial policing were still applicable, and when a revolt in Madagascar in 1947–8 was put down in much the same way, with an estimated 60,000 people killed, there seemed little reason to doubt the French ability to deal with any trouble that might arise. Their response, as the casualty figures imply, was to meet force with force, although it would be unfair to describe it in quite such crude terms. French colonial policy had always been based upon a belief in the civilising mission of the Army as representative and spearhead of European values and control. The result was a policy of 'pacification', developed through the colonial wars of the mid-nineteenth century — particularly those in Algeria in the 1830s and 1840s — which emphasised the need for social and political as well as purely military campaigns. Thus, although the Army might use its superior technology and discipline to defeat any armed threat — characteristics which, inevitably, caused heavy casualties among the dissidents — it would also be used to establish and protect an infrastructure of control, with such benefits as education, medicine and improved agriculture automatically included. The ordinary people of the state would, it was argued, see the advantages of the new rule and cease to cause any trouble. In short, the Army was both the symbol of French power and the instrument of colonial rule.[2]

In itself, this was a sensible, if sometimes rather bloody, set of

Map 2.1: The French in Indochina

principles, little different to those adopted by other colonial powers, and had as its key one of the most important concepts of what would now be termed counter-insurgency — namely the need to combine military and political actions, with emphasis in the long term upon the latter. But in reality, pacification contained within it the seeds of its own destruction. Because of the record of successful application in a wide variety of geographical and political circumstances — in North Africa, Equatorial and West Africa and even Indochina — few people queried its value, little realising its inherent deficiencies. Chief among these was the fact that it was invariably put into practice in conditions favourable to the French — against rebel groups often weakened by religious or tribal divides and poorly armed, organised and equipped. Moreover, in the absence of accurate, on-the-spot reporting, the level of public awareness was usually low and when disquiet was expressed, espcially over disproportionate casualties, it tended to be overshadowed by a widespread, rather vague belief in the civilising mission of colonial expansion. This led, inevitably, to a degree of military complacency, with the tactics of pacification — the establishment of fortified posts in remote areas, designed to dominate and intimidate the local people, backed by mobile 'flying columns' of well-armed troops who would be ready to rush to any trouble-spot and put down the rebels using maximum force — being accepted virtually without question, producing a stereotyped response which was not always appropriate. This became increasingly apparent between 1946 and 1954 in the area of Indochina, comprising the modern-day states of Vietnam (traditionally composed of the three kingdoms of Cochin China, Annam and Tonkin), Laos and Kampuchea (Cambodia).

The French had first shown interest in this region in 1862, when they occupied the eastern provinces of Cochin China, and this was followed a year later by the proclamation of a protectorate over Cambodia. In 1867 the rest of Cochin China was seized and in 1883, after military operations, both Annam and Tonkin followed suit. The new Far Eastern empire was completed in 1893 when Laos also became a protectorate. Isolated rebel movements were encountered, only to be contained and crushed using the traditional methods of pacification, and it was not until 1930, when the Vietnamese Communist Party was formed, that coherence and organisation began to be imposed upon the dissident groups, although the effects were not immediately apparent. The situation changed in 1940, when metropolitan France fell to the Germans, for although all French colonies theoretically remained under the authority of the collaborationist government in Vichy, in reality they were isolated and vulnerable to calls from elsewhere to shift allegiance or bow to particular pressures. In the case of Indochina, cut off from the mainstream of the Second World War and poorly protected by a garrison which could be neither relieved nor reinforced from home, the pressure came from the Japanese, initially to close the supply route from Haiphong to southern China and then to allow the stationing of troops on Indochinese soil. The French colonial administration remained, but was obviously discredited, and when resistance to Japanese presence developed, it did so among the indigenous people rather than the Europeans. In Vietnam, nationalist groups came together in May 1941 to form the *Viet Nam Doc Lap Dong Minh Hoi* ('League for the Independence of Vietnam' — usually shortened to Viet Minh), centred upon the communists under the Annamese Ho Chi Minh, and instigated a guerrilla campaign which was aimed at all aspects of foreign domination. Aided by the Americans after 1943,[3] the Viet Minh gradually carved out bases in the remoter areas of northern Tonkin and gained a large measure of popular support.

This success, coupled to the fact that in March 1945 the Japanese took over the administration of Indochina, detaining or killing the French, left the Viet Minh in a very strong position to exploit the political vacuum which occurred six months later when the Japanese surrendered. On 2 September, Ho Chi Minh declared the creation of an independent Republic of Vietnam in Hanoi while other nationalists took over in Saigon. Unfortunately the new governments were doomed to failure, for the Allies had already decided to put troops into Indochina, ostensibly to disarm and repatriate the Japanese. In the final months of 1945, the British 20th Indian Division landed in the south while Nationalist Chinese forces marched into the north, and although Ho Chi Minh's government was initially untouched, his allies in the south were soon ousted and, by February 1946, the French had reasserted their presence up to the 16th parallel and had re-entered both Laos and Cambodia. There followed a period of negotiation to ensure the withdrawal of the Chinese, during which Ho Chi Minh was forced to co-operate with the French to exert political pressure on Peking, and for a time, as the Vietnamese leader travelled to Paris and openly discussed the idea of independence for his country, it seemed as if a peaceful solution would be found.

But the French had no intention of losing their influence. Their offers of conditional independence for Vietnam were half-hearted and swiftly rejected and, as French troops moved back into the north, reoccupying former garrisons and gradually ousting the Viet Minh, Ho Chi Minh and his military adviser Vo Nguyen Giap decided to force the issue. In December 1946 they mounted an open attack on French positions, only to be defeated by superior firepower in the best traditions of pacification.[4]

As the Viet Minh melted away, the French celebrated what appeared to be a decisive victory. But their euphoria was misplaced, for they were no longer facing ill-organised colonial rebels. The Viet Minh, communist-inspired and strongly national-istic, were disciplined and dedicated revolutionaries, following the pattern of politico-military action currently being perfected in China by Mao Tse-tung. Co-ordinated by a central politburo, the insurgents had already established an infrastucture of control in the rural areas of northern Vietnam, building up 'safe bases' to which they could retire and from which they could sustain a military campaign against the colonial authorities. The overall aim was to gain political control of the state and this was to be achieved by mobilising the support of the people, wearing the French Army down in an attritional guerrilla war and, finally, winning an open conventional battle which would clear the way to a political take-over. Thus the attacks of late 1946 were only part of a far wider strategy and, when they failed, the Viet Minh suffered no more than a set-back. Withdrawing into the 'safe bases' of northeast Vietnam (the 'Viet Bac') and the southern part of the Red River Delta, Ho Chi Minh and Giap concentrated on rebuilding their military forces and strengthening their popular appeal, preparatory to a guerrilla campaign.[5] For nearly three years the French, convinced that the rebels had been defeated, did little to interfere, despite the success of a similar campaign in China. Indeed, when Mao emerged victorious in October 1949, the French position in Indochina was already compromised, for once supplies began to flow into the Viet Bac from China, the strategic balance tilted decisively. If the French concentrated upon interdicting such supply routes, they would denude defences elsewhere; if they ignored their existence, they would face the prospect of virtually unlimited Viet Minh strength.

Such a dilemma indicates the scope of French problems, for pacification took no account of the subtleties of communist revolutionary warfare. When the Viet Minh reopened their attack in late 1949, therefore, the French were caught largely unawares and badly deployed. Their isolated garrisons, often consisting of no more than a squad of soldiers under a junior officer or NCO, were extremely vulnerable to guerrilla assault and casualties were high. Even where the garrisons were more substantial, as along the Cao Bang-Lang Son ridge in northeast Vietnam, the French found it impossible to maintain their positions, particularly when the supply of such garrisons depended upon motorised transport through difficult terrain. The convoys using *Route Coloniale* 4, linking Lang Son and Cao Bang, were particularly vulnerable to guerrilla ambush and, as losses mounted, the morale of the French forces suffered. Thus, when the Viet Minh switched to direct attacks on the ridge garrisons in 1950, they had already established a moral ascendancy which soon produced military victory. In September the outpost of Dong Khe was captured, forcing the French to evacuate Cao Bang in the north and, as a column of soldiers and camp followers took to the road, they came under constant attack. A parachute battalion — representing the mobile element of pacification tactics — was hastily committed, but merely acted as a focus for Viet Minh attentions and was virtually wiped out. By the end of October the French had lost over 6000 casualties and had been driven out of the northeast of the country, leaving it to be developed as the strongest of the communist bases.

Such a disaster should have led to a reappraisal of strategy, but events precluded this. Imagining that he had the French on the run, Giap mobilised his forces for an all-out attack on the Red River Delta, hoping to defeat his enemy in open battle and take Hanoi. It was a premature move, for the French government had responded to the Cao Bang débâcle by despatching substantial reinforcements to Vietnam and appointing General Jean de Lattre de Tassigny as both Commander-in-Chief and Governor-General. He immediately withdrew all remaining garrisons into the delta region, set up outposts on the approaches to Hanoi and

Haiphong and reorganised the elite units of his forces — the paras, Marines and Foreign Legion — into effective mobile groups, backed by air power, artillery and armour. Thus, when Giap carried out his attacks in 1951 — in the northeast at Vinh Yen (13–17 January), in the north at Mao Khe (23–28 March) and in the south around Phat Diem (29 May–18 June) — he was badly defeated, losing an estimated 12,000 casualties to the superior firepower of the French. This seemed to vindicate the basic principles of pacification and set the pattern for future French actions. An elaborate defence system — the de Lattre Line — was set up around the delta to protect the French base and the emphasis was placed firmly upon the technological advantages of artillery and air power. In late 1951, de Lattre took the process one stage further, sending paratroops to capture Hoa Binh, 40 km (25 miles) outside the delta defences, and consolidating the new position with mobile groups which linked up along the roads and rivers.

This was just what Giap wanted, for his response to the defeats of 1951 had not been to give up the fight but to revert to guerrilla warfare in accordance with the strategy of revolution. Thus, as the French pushed out from their base, their supply lines were attacked and the garrison at Hoa Binh isolated, forcing de Lattre's successor, General Raoul Salan, to order a withdrawal. By February 1952 the French were back in the delta and Giap had recovered the military initiative. A similar French move in October 1952 – Operation 'Lorraine' — in which Salan used the bulk of his mobile reserve in a pincer attack towards Viet Minh supply dumps at Phu Tho and Phu Doan, north of the Red River, also failed in the face of determined guerrilla attacks, and by early 1953 the colonial forces were running short of options. A Viet Minh invasion of northern Laos in April 1953 forced Salan to commit a substantial part of his army to 'hedgehog' positions on the Plain of Jars, further weakening his capabilities and undermining his offensive strength. To all intents and purposes, pacification had failed.[6]

But this was not immediately apparent. In November and December 1952 a garrison at Na San, close to the Laotian border, had held out against wave after wave of Viet Minh attackers from positions which were sustained exclusively from the air. This suggested a potential for success which, if exploited elsewhere in Vietnam, would force the Viet Minh into open battles on ground of French choosing and subject to the full strength of French technology. As a result, in November 1953 General Henri Navarre, who had replaced Salan in the previous May, authorised Operation 'Castor' — the dropping of parachute forces into the valley of Dien Bien Phu, 275 km (170 miles) to the west of Hanoi, where the establishment of a centre of resistance on the pattern of Na San would interdict Viet Minh supply lines into northern Laos. Initially the aim was to create an anti-guerrilla 'mooring-point' — a base from which French-officered T'ai forces could attack Viet Minh targets in northwest Vietnam — but once Giap reacted by committing his regular formations to oppose French consolidation, Navarre ordered the valley to be fortified and equipped to fight the set-piece battle he was convinced would destroy the rebel movement. Unfortunately, French intelligence was poor, Dien Bien Phu was isolated and dependent upon air supply which could not always be guaranteed (especially after the arrival of Red Chinese anti-aircraft units in the hills overlooking the valley) and the concentration of overwhelming Viet Minh strength took the colonial authorities by surprise. In January 1954 the base — consisting of a number of fortified strongpoints around an airstrip and central command bunker — came under artillery fire from the surrounding hills, the capture of which had not been considered important by the French commanders, and in March outlying posts to the north — codenamed 'Anne Marie', 'Gabrielle' and 'Beatrice' — fell to determined Viet Minh attack or were abandoned. The airstrip was closed to traffic, forcing the French to depend upon parachute resupply which was never accurate enough to sustain the rapidly shrinking perimeter, and after a siege conducted along the traditional lines of trench warfare, the central command bunker fell on 7 May. Over 7000 French troops — most of them elite parachute and Foreign Legion personnel — had been killed and nearly 11,000 captured, and although the Viet Minh lost an estimated 20,000 men, their victory was decisive.[7] It coincided with a similar French failure around Tourane (Danang) in Operation 'Atlante'. Landings had been made on 20 January 1954, but after initial success an inevitable Viet Minh counter-attack had forced Navarre to commit the last of his strategic reserve. By May the French were militarily bankrupt.

Any form of military solution was now precluded. The

government of Pierre Mendès-France was obliged to bow before an outburst of popular fury and negotiate an end to French presence in Indochina. The last troopship sailed from Saigon in November, leaving Vietnam divided, rather arbitrarily, on the 17th parallel and both Laos and Cambodia as independent states.

Such a general account of the war is designed to show how pacification was no longer a viable response to the new patterns of insurgency which emerged in the aftermath of the Second World War and how the twin elements of static garrisons and mobile columns were gradually undermined and destroyed in a conflict which bore little similarity to previous colonial revolts. In itself this may be a reasonable analysis of the process of French defeat, but there was more to it than that. In terms of counter-insurgency — the destruction of an internal threat to the existing administration of a state — the French displayed a number of crucial weaknesses, not all of which can be excused by the fact that they were the first of the European powers to face a closely-integrated communist revolution organised along Maoist lines.

There was, first and foremost, no official recognition that the threat was essentially a political one, with the result that throughout the war little attempt was made by the French to counter Viet Minh efforts to gain popular support. What propaganda there was lacked effect or relevance to the cause of Vietnamese nationalism, being based almost exclusively upon the premise of continued colonial dominance, while the record of French defeat both at home and in Indochina between 1940 and 1945 gave the Viet Minh a central theme of counter-propaganda which they did not hesitate to exploit. At the same time, French promises of independence were discredited by the failure of the Paris negotiations of 1947–9 and further undermined once it became apparent that any such promises lacked the backing of sufficient armed force to ensure their implementation. Political instability in France itself led to constant changes of government, precluding the evolution of consistent policies or aims, and the fact that the war, coming so soon after the traumas of 1940–5, was intensely unpopular at home inevitably affected its conduct in Indochina. Politicians desperate for domestic support did little to ensure the commitment of adequate resources — in 1950, for example, as Giap's forces were taking the Cao Bang ridge, the size of the French Army in Indochina was ordered to be reduced by 9000 men and an amendment to the Budget Law restricted the use of conscripts to 'homeland' territory only (France, Algeria and the occupation zones of Germany) — while the substantial Communist Party in France took every opportunity to hamper military efficiency, even to the extent of sabotaging equipment awaiting shipment to the Far East.

Such a lack of political direction was reflected in the command structure of the forces in Indochina, for although the appointment of de Lattre to positions of both civil and military authority in late 1950 implied a co-ordinated approach, similar to that being imposed by the British in Malaya at much the same time, in reality he enjoyed little initiative and could make few decisions without reference to his political masters at home. When de Lattre resigned through ill-health a year later, the experiment was not repeated with his successors, leaving both Salan and Navarre subject to interference not just from Paris but also from the civilian Governor-General in Vietnam.

Even if this had been a satisfactory arrangement — after all, in successful counter-insurgency there should be a degree of political control — it is highly unlikely that a different military outcome would have resulted. The French, despite valiant efforts to defeat what they saw as a rebel movement, totally misinterpreted the nature of the threat. Their intelligence was poor, concentrating almost entirely on the regular formations of the Viet Minh and even then failing to provide a complete picture — at Dien Bien Phu the defenders of the valley were assured that the enemy had no anti-aircraft capability, yet a total of 62 planes were later shot down or badly damaged. Information about the organisation and strength of guerrilla forces and, more importantly, the political infrastructure of the Communist Party, was lacking, and no serious attempts were made to rectify this — there were, for example, no policies designed to tempt deserters from the Viet Minh or to recruit captured guerrillas into 'pseudo-gangs' for deployment against their erstwhile colleagues. Some local tribesmen — principally from the T'ai mountains in the north — were organised, under French officers and NCOs, into *Groupements de Commandos Mixtes Aéroportés* (GCMA — renamed *Groupements Mixtes d'Intervention* (GMI) in December 1953) for anti-guerrilla operations in remote areas, but they were generally distrusted and rarely co-ordinated with regular units. The French even showed a marked reluctance to raise a national army in Vietnam, preferring to recruit loyal Vietnamese into regular French units — for example, when the 2nd Battalion, 1st Parachute Light Infantry jumped into Dien Bien Phu on 20 November 1953, nearly half its number were Vietnamese[8] — and refused adequately to arm local village militias, the existence of which should have relieved the regulars of the need to man static defences. When it is added that few French troops displayed either a willingness or an aptitude for unconventional warfare, the inappropriate nature of the response may be fully appreciated.

Nor did the shortcomings end there, for the general dependence upon technology and firepower, although occasionally an asset, too often worked against the French. Pacification tactics demanded an element of mobility, yet this was never achieved to an extent which could prove decisive. During the Indochina War there were never more than ten helicopters available, none of which could be used for the transportation of large bodies of troops, so the French had to depend upon parachute forces, backed by columns of men in half-tracks, tanks and trucks. The paras were certainly effective fighting soldiers, but there was never sufficient airlift to ensure unlimited deployment potential or guaranteed resupply: by the time of Dien Bien Phu, Air France passenger and freight aircraft, together with their crews, were having to be requisitioned in a desperate attempt to keep the supply corridor open, with predictably poor results. On the ground, the mobile columns stuck to the roads, offering ideal ambush targets, while their supply back-up was always vulnerable to interdiction. During Operation 'Lorraine' in late 1952, for example, the advance beyond the Red River Delta had to be curtailed because of over-extended supply lines and when a withdrawal was ordered, the Viet Minh disrupted it with ease: on 17 November a particularly effective ambush in the Chan Muong gorge left a tank and six half-tracks destroyed and over 300 French dead or wounded. In many ways, it was a symbol of the failure of French techniques.

It would be wrong to dismiss the Indochina War in entirely negative terms, however, for some basic counter-insurgency lessons did emerge. It was recognised, for example, that the guerrillas gained considerable strength from their support among the local population, and although it was never adopted as a general policy in Vietnam, valuable experiments in resettlement were conducted in the border areas of Cambodia in 1946 and

1951. The aim was to remove the sources of food, shelter and recruits from the Viet Minh, and some success was enjoyed. Similarly, the operations carried out by the GCMA/GMI, although afforded little support or publicity, did show what might be achieved using anti-guerrilla forces, capable of meeting the enemy on his own ground and with his own tactics. But these were isolated developments, more important in retrospect than at the time: in the end the French were outmanoeuvred and outfought in a war to which they could only respond with inappropriate, archaic tactics.[9]

The scale of the defeat in Vietnam had profound effects upon the French Army, acting as a catalyst for the evolution of *guerre révolutionnaire*. This had its origins among officers who had fought in Indochina and especially among those who had been captured by the Viet Minh, learning at first hand the nature of the communist revolutionary process. Men such as Colonels Roger Trinquier and Charles Lacheroy (both of whom wrote articles and books in the mid-1950s, disseminating their views)[10] approached their analysis of the defeat on the basis of two key premises. First, they believed that Indochina had been merely a part of a worldwide communist conspiracy and that the French Army had been acting, without either domestic or international backing, as the sole defender of the West and its values. Second, they realised that the threat in Indochina had not come from an ill-organised, poorly co-ordinated rebel movement, but from a highly structured and fully integrated group of dedicated revolutionaries, intent on the overthrow of existing political structures using a unique mixture of psychological and military means. Indeed, the theorist Colonel Georges Bonnet went so far as to synthesise the process into an equation: 'partisan (guerrilla) warfare + psychological warfare = revolutionary warfare'.[11]

This was undoubtedly an oversimplification, but it does indicate the extent of the analysis taking place. After 1954 French officers avidly read and absorbed all they could find on the principles of Mao's revolutionary process, even constructing elaborate models of the ways in which it might be carried out. Basing their views on the examples of China and Vietnam, they discovered a series of distinct 'phases' in the evolution of the communist campaign, starting with the infiltration of the population by political cadres, progressing through the creation of a guerrilla infrastructure and alternative government system, and culminating in a co-ordinated, all-out offensive against the existing authorities, aimed at seizing political power. Within such phases, the importance of popular support, international backing and the steady demoralisation of government forces was highlighted and stressed. It was summed up neatly by Commander J. Hogard in his article '*Guerre révolutionnaire et pacification*', published in 1957:

> . . . revolutionary warfare is very different from traditional conventional war. Widely dispersed at the outset, it gradually draws strength and resources from the enemy, seeking to capture not military or geographic objectives, but the population. When the situation is ripe it concludes with a single all-out effort in which all its forces are concentrated — if this is still necessary. Generally, the last battle is won before it has started since the enemy is swamped, groggy, demoralised — he is psychologically ready to be defeated.[12]

But this was only half the picture, for once the nature of the threat had been recognised, it was only logical for the theorists to suggest a doctrine of counter-revolution. They began by pointing out what they felt were the inherent weaknesses of the Maoist model — its vulnerability during the initial stages, before deep-rooted support among the people had been established, its dependence upon a logistic base, often in a neighbouring country, its military inferiority during most of the evolutionary process — and went on to compile a series of counter-measures designed to exploit these factors. It was essential, they argued, that governments should maintain a careful watch over their people, particularly in remote areas, adopting policies of education, reform and firm military action to prevent or contain communist subversion. If this proved impossible — if, for example, the danger was not recognised until after subversion had begun — then even more drastic measures should be carried out without hesitation or hindrance, centred upon the need to cut the insurgents off from their sources of support among the people and beyond the borders of the state. Extensive resettlement of vulnerable groups within the population and the building of elaborate barriers along international frontiers were just two of the suggested policies, but it was stressed throughout that government support for such measures, however unpopular or repressive they might seem, had to be absolute. The Army should never operate in a vacuum, as it had appeared to do in Indochina: it had to be fully supported by a domestic government which appreciated the danger and by international opinion, at least in the West, which understood the need for violence. In short, to defeat a politico-military conspiracy, the French had to be prepared to adopt a politico-military counter-doctrine, based upon an ideological strength of purpose equal to that of the communists. It was a tall order, fraught with dangers of military politicisation and inflexible, almost blinkered, preconceptions.

It would be wrong to imagine that the French army entered the Algerian War in 1954 with the theory of *guerre révolutionnaire* already worked out. Its evolution took time — most of the key articles and books did not appear until 1956–7 — and it was perhaps unfortunate that an insurgency occurred before the process was complete. Inevitably, as many of the theorists were serving officers and involved in countering the new campaign, Algeria became a testing-ground in which policies were advocated before they had been fully analysed, or put into effect without the necessary political understanding or support. More importantly, Algeria did not fit the pattern of revolution so laboriously compiled, reflecting nationalist demands for independence rather than the latest stage in a global communist conspiracy. In such circumstances, *guerre révolutionnaire* had all the ingredients of disaster.

The Algerian War, which began on 1 November 1954 with a series of badly co-ordinated bomb attacks against French targets, was a confused affair. The main insurgency group — the *Front de Libération Nationale* (FLN), created from within the plethora of anti-French political parties which had evolved in the early 1950s — was dedicated to a policy of guerrilla warfare and terrorism which would force the French to grant independence and withdraw. But Algeria was never a colony — since December 1848, some 18 years after the French arrived in North Africa, it had been an integral part of metropolitan France, electing representatives to central government and enjoying the same status as Normandy or Provence — and although the majority Moslem population had many of the characteristics of a subject people, having been ousted from the best land and denied their rightful say in local politics by French settlers (*colons*, or more graphically, *pied noir*), their demands were clearly impossible to

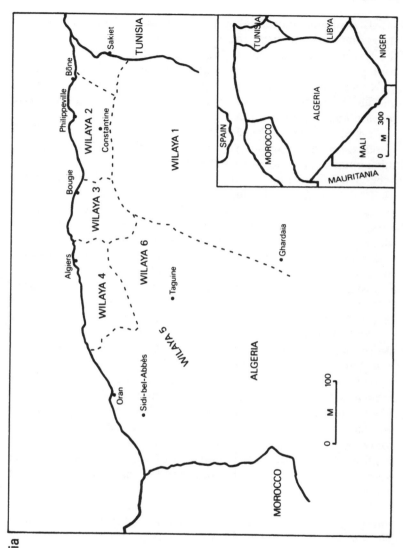

Map 2.2: Algeria

satisfy without undermining the structure of France itself or triggering violence from the notoriously anti-Moslem *colons*. In the event, the war soon degenerated into a three-cornered conflict, involving the French government, the *colons* and the FLN, with the Army occupying an unenviable position in the middle.[13] The strains, made worse by the shortcomings of *guerre révolutionnaire*, were intolerable.

At first, however, the French did enjoy some success. Their response to the bomb attacks of 1 November 1954 was familiar: substantial armed forces were deployed to the affected areas and, backed by predictable *colon* violence, they swiftly crushed the FLN, dispersing its leadership and breaking up the guerrilla gangs. This was followed, in the traditions of pacification, by attempts to reform local government in Algeria, increasing Moslem participation, but the French advantage was short-lived. *Colon* mistrust undermined the reforms, while the FLN, despite (or perhaps because of) its failure, attracted fresh recruits. In an atmosphere of nascent Moslem nationalism, affecting states throughout the Middle East and North Africa, the Algerian rebels began to receive external support, increasing their demands as the French granted independence to the neighbouring protectorates of Morocco and Tunisia in 1956 and setting up 'safe bases' across the new frontiers. With a revival of guerrilla activity — manifested most dramatically in August 1955 when 123 French settlers were murdered in Philippeville (an incident which, once again, led to *colon* violence) — it began to seem as if the pattern of Vietnam was about to be repeated. A rebel conference, held in the Soummam valley in Kabylia in mid-1956, reinforced this view as the remnants of the old FLN leadership were ousted and replaced by more militant men under Ramdane Adane.

The new rebel campaign concentrated upon urban terrorism within Algiers, where FLN networks, using the rabbit-warren of Moslem streets in the Casbah as a refuge, initiated a series of shootings and bomb attacks in 1956. The aim appears to have been to provoke a violent response from the *colons*, thus driving a wedge between the moderate Moslems and the French, and support for the FLN was further ensured by a widespread use of intimidation within the Casbah. By the end of the year, with *colon* vigilantes exacting revenge on Moslems throughout the city, Algiers was on the brink of anarchy. On 7 January 1957 the

French Resident-Minister, Robert Lacoste, ordered General Jacques Massu, commander of the 10th Colonial Parachute Division, to assume full powers in the city and root out the terrorist teams. It was the chance that the theorists of *guerre révolutionnaire*, many of whom were serving under Massu's command, had been waiting for. They began to try out their ideas, dangerously free from political control, without delay.

Their first priority was to impose military control, particularly in the Casbah, where Colonel Marcel Bigeard's 3rd Colonial Parachute Regiment was deployed. This was soon achieved through a tough policy of house-to-house searches, constant foot patrols and identification checkpoints. An attempted general strike by the Moslems was ruthlessly crushed on 28/29 January and by the end of the month the violence had subsided. But the FLN networks had merely disappeared deeper underground, requiring the collection and collation of detailed intelligence to isolate and destroy them once and for all. One of Massu's first acts on entering the city had been to seize all police files; when these were analysed, large numbers of arrests were ordered and the suspects handed over to a special *Détachement Opérationnel de Protection* for interrogation. This was invariably brutal and often involved the use of torture, but the results were impressive.

Information thus gathered was used by Massu's chief of staff, Colonel Yves Godard (an advocate of *guerre révolutionnaire* and a firm believer in *Algérie française*) to build up a picture of the FLN framework. Beginning with a blank order of battle known as an *organigramme*, he gradually added names as they became apparent, until he had before him a detailed chart which could be used as the basis for further arrests. At the same time, Colonel Roger Trinquier, another *guerre révolutionnaire* specialist, set up a more covert intelligence-gathering network — the *Dispositif de Protection Urbaine* — in which Algiers was divided into sectors, sub-sectors, blocks and buildings, each containing senior inhabitants (usually Moslem ex-soldiers still loyal to France) who would act as 'spies', reporting suspicious movements and keeping a general check on the activities of local people.

By March 1957 this combination of aggression and information had succeeded in breaking up the FLN infrastructure, but the crisis was not yet over. The hard-core rebel leadership, under Saddi Yacef, was still at large, and although the paras were withdrawn at the end of March, they had to be re-committed, on the same *carte blanche* terms, when a new wave of bomb attacks began in June. This time Massu was determined to destroy the FLN, adopting the by now familiar tactics of patrolling, intelligence-gathering, arrests and torture but with even more aggression. By late September Yacef had been captured and two weeks later the last FLN stronghold, deep in the Casbah, was taken out.[14]

In purely military terms, Massu's campaign could not be faulted. He had carried out his orders efficiently, freeing Algiers from the violence of terrorist activity and destroying the FLN urban network. But politically the campaign was a disaster for the French. As the full scale of the paras' actions became apparent — altogether some 3000 Moslems, including the FLN leader Ben M'hidi, had died or 'disappeared' while in detention — and as reports of torture seeped out, many people began to express their revulsion at the methods employed. In Algeria all hopes of mobilising moderate Moslems onto the side of the French disappeared (as the FLN had intended); in France public opinion began to swing in favour of a political settlement which would eventually lead to Algerian independence; in the wider world France stood condemned for adopting policies which were, in many eyes, little different to those of the Gestapo or SS. Army actions in Algeria were generally discredited, regardless of future events, for the 'Battle of Algiers' had driven a permanent wedge between the political and military aspects of counter-insurgency.

This was unfortunate, for elsewhere in Algeria — particularly in the rural areas — the methods used by the French Army enjoyed a remarkable degree of success without recourse to unethical tactics. Based upon the 'lessons' of Indochina and the analysis of *guerre révolutionnaire*, the first priority was seen to be the isolation of the guerrillas from outside support. This was achieved by the construction of physical barriers (*barrages*) along the borders of both Morocco and Tunisia once these states had gained independence and declared their support for the FLN. Of the two, Tunisia was the more active, playing host to the Algerian 'army-in-exile' and acting as supply base for the guerrillas in the field, and the French reacted by building a elaborate barrier — the Morice Line — down the full length of the frontier. Comprising an electrified fence, surrounded by minefields and constantly patrolled, the line was completed in September 1957; thereafter FLN groups tried every conceivable means of breaching the wire, using high-tension cutters, Bangalore torpedoes, tunnels and ramps, but to little effect. Any guerrillas who did get through were invariably picked off by Army patrols, backed by armour, artillery, air power and heliborne quick-reaction units. By April 1958, after the destruction of a particulrly large penetration group round Souk-Ahras, the kill-rate of infiltrators was reported to be as high as 85 per cent, after which few attempts were made to breach the line or, indeed, its less elaborate counterpart on the Moroccan border. More importantly, the process split the rebel movement, isolating the guerrillas of the 'interior' from the mainstream of nationalist politics in Tunisia, where new leaders emerged who were more interested in creating a regular army for intervention in Algeria to ensure FLN rule once independence had been granted than in a continuation of low-level insurgency.

The next logical step for the French was therefore to destroy the guerrilla networks of the interior, and this they did by adopting many of the techniques associated with successful counter-insurgency. Large numbers of Moslems living in areas of known or suspected guerrilla activity were physically removed — over 300,000 from around Constantine alone — and resettled in 'model' villages where intelligence teams operated. At the same time a determined psychological warfare campaign was mounted to persuade the Moslems that French rule was an attractive long-term proposition. In 1956 Colonel Charles Lacheroy — yet another of the *guerre révolutionnaire* activists — set up a Psychological Action and Information Service which, by the following year, had been regularised as the 5th Bureau attached to every headquarters and regional command in Algeria. Run by members of the Special Administrative Section (SAS), the aim was to destroy enemy morale while boosting that of the French, and although the SAS became over-politicised, leading to its disbandment in early 1960, some success was achieved.

This was exploited by effective and appropriate military tactics, based upon a system known as *quadrillage*. In many ways this differed little from the previous tactics of pacification, in that the country was covered by a chequerboard of small garrisons, backed by mobile forces, but new refinements had been added since Indochina. There was a greater emphasis on intelligence, much more attention was paid to 'hearts and minds' (the normal procedure tended to involve the deployment of conscripts to the small garrisons on the presumption that such 'civilians in uniform' would be better able to relate to the local people) and the whole thing was backed by large-scale, well co-ordinated sweeps of selected areas. Known as *ratissage*, these sweeps proved effective against demoralised and weakened guerrilla groups, the most impressive being carried out in 1959 under the general title of the Challe Offensive. Organised by General Maurice Challe, newly-appointed Commander-in-Chief, the attacks followed a set pattern. In an effort to pinpoint the guerrilla gangs, pro-French Moslem units (*Harkis*) would move into an area of known FLN activity, living off the land and playing the guerrillas at their own game. Once the FLN groups had been discovered, mobile *Commandos de Chasse* would step in, pursuing the enemy preparatory to the kill. This would be carried out by the elite units of the *Réserve Générale* — paras, Marines and Foreign Legion, backed by armour, artillery and air power. The first operation took place in February 1959 in the Ouarsenis mountains to the southeast of Oran and was followed between June and October by a major sweep into the Kabylia range to the southeast of Algiers. By the end of that period, the FLN had virtually ceased to exist as a guerrilla force.[15]

In a conventional war, this degree of destruction would have constituted a total victory, but in the context of a politically-motivated insurgency it proved to be almost irrelevant. By 1959 irreparable damage had been done to relations between French military and political personnel by the revelations of the Battle of Algiers, and no politician was prepared to back the Army with equally tough measures to ensure a continuance of French rule. This led inevitably to frustration in the Army, particularly among those officers who believed in the conspiracy theory at the core of *guerre révolutionnaire*. To them, the key to success had to be a firm belief in the role of both Army and politicians as protectors of the West against communist revolution; when the politicians clearly failed to appreciate this, the only answer seemed to be an assumption by the Army of responsibility for political decisions. As Colonel Antoine Argoud, one of Massu's staff officers, put it in 1960: 'We want to halt the decadence of the West and the march of communism. That is . . . the duty of the army. That is why we must win in Algeria.'[16]

The results were predictable, but none the less tragic. By 1958, convinced that successive French governments were blind to the true danger, Generals Salan and Massu openly associated themselves with the rising tide of *colon* disaffection, joining a Committee of Public Safety in Algiers and threatening to intervene with force in Paris unless Charles de Gaulle — a man of sufficient strength of purpose, it was felt, to understand the Army's point of view — was invited to form a government. Unfortunately, although de Gaulle did enter office, he proved to be unimpressed with the basic ideals of *guerre révolutionnaire* and, faced with the intractability of the Algerian problem, began to veer towards negotiation with the exiled FLN leadership in Tunisia, leading to eventual independence. In response, elements of the Army — particularly the paras, among whom so many of the *guerre révolutionnaire* theorists were to be found — did nothing to quell *colon* disturbances in early 1960 ('the Week of the Barricades') and, in April 1961, actively supported an attempted coup led by Salan and three other general officers. It was a logical culmination of *guerre révolutionnaire*: if the politicians refused to understand or believe in the theory, then they had to be replaced.

In the event, de Gaulle was able to rally the bulk of the armed forces and the vast majority of the people of France, but the damage was crippling. Once military motives and policies were thus discredited, counter-insurgency could never succeed. De Gaulle negotiated independence — effective from the agreement at Evian in July 1962 — and disbanded those elements of the Army which had proved disloyal. Officers were put on trial, while others went underground to form the *Organisation Armée Secrète* (OAS), conducting their own brand of terrorism against Moslems and those Frenchmen who supported withdrawal, but the end result was never in serious doubt. Despite a remarkable record of military success and the evolution of sophisticated techniques of counter-insurgency, the political nature of *guerre révolutionnaire* had all but destroyed the reputation and effectiveness of the Army. Rigid preconceptions had led to drastic actions, taking elements of the Army out of their normal position of subordination in a democratic society and plunging them into the dangerous waters of overt interference in the domestic politics of the state.

Since 1962 little has been heard of *guerre révolutionnaire*, even though some of its techniques were undoubtedly valid. Instead, the French Army has been largely kept out of counter-insurgency situations, being used in a far more flexible and politically controlled role as a *force d'intervention*. Based in southern France, with a forward element on constant alert in Senegal, this emerged as early as 1964, organised around the 11th Airborne Division with its paratroop, marine, infantry, artillery and seaborne units, and was dedicated to the protection of French commercial, economic and strategic interests throughout black Africa. Most of the French colonies in Africa had received their independence by 1960, but influence had been maintained through policies of economic aid, political support and military agreements, the latter often providing base facilities for the French in exchange for promises of force deployment in the event of external or internal threats to the new post-colonial governments. According to French sources, such deployments were regularly carried out in the early 1960s, as Chad, Cameroon, Congo-Brazzaville, Mauritania, Niger and Gabon all requested assistance.[17]

This set the pattern for future policy and since then interventions have continued almost as a matter of course — in Chad (1969, 1978 and 1983), Western Sahara (1975, in support of Mauritanian claims over this disputed territory), Zaire (1977 and 1978, to protect or rescue European hostages in the eastern province of Shaba) and the Central African Empire/Republic (1979, to overthrow the notoriously unstable and politically embarrassing Emperor Bokassa) — but in each case French forces have tended to remain aloof from the detailed politics of the states involved, preferring to gain or protect the required level of influence before withdrawing. Even where anti-French insurgents are clearly operating to destabilise the politics of a particular state, there is little evidence to suggest the adoption of counter-insurgency policies, with the result that long-term or permanent solutions to the problems which initially precipitated intervention rarely emerge. Nowhere is this more true than in Chad, a vast central African country of some 1,284,000 sq.km (386,209 sq. miles), rich in the sort of mineral deposits which attract outside interference.

Chad received its independence from France on 11 August 1960, but was immediately plunged into the chaos of civil war. Although the population was small — about 2.5 million — it was multiracial, multilingual and deeply divided along religious lines. The French had handed power to the black Africans of the south,

Map 2.3: Chad

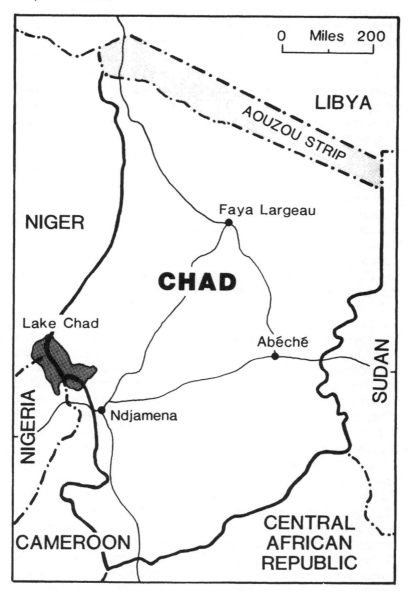

led by François Tombalbaye, but his rule was opposed by
northern tribesmen who were mainly Arab-speaking and
Moslem: factors which guaranteed them at least a degree of
support from the neighbouring states of Sudan, Libya and
Nigeria. Tombalbaye soon found that the all-party political
framework left by the French was unworkable, and in 1963 he
dismissed all Arab-speaking ministers from his cabinet, dissolved
all political parties except his own and introduced harsh anti-
tribal laws aimed specifically at the northerners. Their reaction
was predictable: led by Abba Siddick, erstwhile Minister of
Education, they formed the *Front de Libération Nationale de
Tchad* (Frolinat) and raised the standard of revolt in the northern
mountains of Tibesti province. By the mid-1960s clashes between
Tombalbaye's *Armée Nationale de Tchad* (ANT) and Frolinat
were increasing in scale and frequency, with unrest spreading to
the east and centre of the country. When the capital Fort Lamy
(later renamed N'Djamena) was threatened in 1969, General
Malloum, ANT chief of staff, persuaded the president to turn to
the French for military aid, invoking the Franco-Chadian treaty
that had been signed at the time of independence.[18]

Because of domestic reservations about the efficacy of aiding
an obviously unstable and undemocratic regime, de Gaulle
responded by despatching little more than a token force — five
companies of the 2nd Foreign Legion Parachute Regiment and a
compagnie de marche, composed of volunteers and men of the
2nd Foreign Legion Infantry Regiment. On arrival in Chad in
September 1969, the troops were divided into two mobile
columns, the organisation and subsequent operations of which
bore more similarity to the old tactics of pacification than to the
counter-insurgency ideas of Algeria. Infantry, mounted in half-
tracks and supported by armoured cars and armed helicopters,
drove into areas of known insurgent activity, tempting the
Frolinat groups into open battles which they could never win.
The fighting was occasionally hard and the size of the French
contingent had to be increased — by early 1970, over 3600
troops, including 900 Legionnaires, 1000 Marines and 1000 Air
Force personnel, were available in Chad[19] — but the rebels stood
little chance against the well-proven combination of mobility and
firepower. Even so, the French made no attempt to build on their
success by introducing policies of civic action, propaganda or
hearts and minds — indeed, the overtly military response to the

crisis by both the ANT and the French Army probably did more to alienate the ordinary people of the north than to solve the chronic social problems of the state and so ensure the future strength of Tombalbaye's government. Frolinat was allowed to retire and recuperate, using bases in Libya, where the advent to power of Colonel Moamar al Gaddafi in September 1969 guaranteed long-term support for the Moslem dissidents. Thus, although the level of violence in Chad had declined, enabling the French to withdraw in December 1971 (having lost eight men killed and 90 wounded), the basic problems of the state had not been addressed and all the causes of trouble remained.

There is evidence that Tombalbaye realised this, putting forward his own rather odd response by forging links with Gaddafi and turning against the French (the small garrison left behind in N'Djamena was abused and attacked) in a desperate attempt to pre-empt the resurgence of rebellion. Unfortunately this created more problems than it solved, producing splits in the ruling party and leading Tombalbaye into even more extreme actions. By 1975, with a 'cultural revolution' (*Mouvement National pour la Révolution Culturelle et Sociale*) in full swing, Malloum and the ANT had seen enough: on 13 April a military coup led to the death of Tombalbaye and the formation of a new regime under Malloum. Almost immediately he faced a fresh wave of insurgency in the north as Frolinat, by now recovered and under the more dynamic leadership of Goukouni Oueddei and Hissène Habré, initiated guerrilla attacks. The pattern of the late 1960s seemed about to be repeated.

At first Malloum appeared to enjoy some success — by early 1978 he had managed to split the Frolinat leadership by offering Habré the post of prime minister in a new, ostensibly multi-party, government — but the increased level of Libyan involvement in the north, where Gaddafi had his eye on the mineral-rich Aouzou Strip, soon persuaded the Chadian junta to request French aid once again. President Valérie Giscard d'Estaing responded by sending a second and larger expeditionary force, this time comprising the whole of the 2nd Foreign Legion Parachute Regiment, a squadron of armoured cars from the 1st Foreign Legion Cavalry Regiment, two companies of the 2nd Foreign Legion Infantry Regiment and a company of *Infanterie de la Marine*, backed by Jaguar and Breguet aircraft as well as armed and troop-carrying helicopters. Given the situation that had existed in 1971, the French were surprised to discover how much territory in the north had already fallen to Frolinat forces, although this did not prevent a repetition of the tactics used in the earlier intervention. Once again, the combination of mobility and firepower achieved spectacular military results — in one particular action around the small town of Ati, to the east of N'Djamena, in May 1978, for example, a column of paras, Marines and armoured cars, supported by missile-carrying Alouette helicopters and Jaguars, managed to destroy a group of over 200 guerrillas in a running fight that took a week in temperatures exceeding 49°C (120°F)[20] — but it was not long before the confused politics of Chad undermined success. By means of convoluted diplomatic manoeuvres, in which the Libyans played a major part, Malloum and Habré accepted the establishment of a *Gouvernement d'Union Nationale de Transition* (GUNT) in November 1979, with Goukouni as president and Habré as minister of defence. In response, the French ceased all military operations in the north and withdrew their forces into N'Djamena to protect French civilians caught in a fresh wave of anti-Gallic frenzy. It was not a situation that could be readily appreciated by the expeditionary force, especially when in March 1980 Goukouni ousted Habré, who fled to Algeria. The French were now in the bizarre position of supporting the man they had arrived in Chad to oppose and the decision to withdraw completely from the country in April 1980 was the only logical one to take.

But the civil war was by no means over. Habré spent the next two years raising his own *Forces Armée du Nord* (FAN), and on 7 June 1982 he staged a successful attack on N'Djamena, forcing Goukouni to flee first to Cameroon and then to Libya, where he put together a GUNT army of opposition. Supported by Gaddafi, whose pretensions towards the Aouzou Strip were now well known, GUNT forces gradually reasserted their previous control over Tibesti and, while Habré was preoccupied by a frontier dispute with Nigeria around Lake Chad, Goukouni struck south. On 24 June 1983 his troops took the oasis town of Faya Largeau, overwhelming men of Habré's *Forces Armées Nationales Tchadiennes* (FANT) with sophisticated Soviet weapons supplied by Gaddafi. Habré called desperately for foreign aid, receiving troops from Zaire and financial backing from both China and the United States, but the French showed a marked (and understandable) reluctance to become too deeply involved. Further heavy fighting around Faya Largeau — the town was recaptured by FANT forces in July, only to be lost again the following month — and evidence of substantial Libyan troop deployment into Tibesti alarmed the Americans, however, and under pressure from them and from moderate black African states, President François Mitterand was persuaded to reconsider. In mid-August a third expeditionary force was committed to Chad, comprising 2800 men of the *Force d'Assistance Rapide* (the modern equivalent of the *force d'intervention*), backed by Jaguar and Mirage jets. Staging through N'Djamena, they advanced as far as the 15th parallel before setting up the 'Red Line', based on the towns of Abéché, Salal, Moussoro and Biltine, to prevent a further push by Goukouni's forces. At first they were under strict orders not to fire unless fired upon — a policy which led them to stand apart from clashes between FANT and GUNT forces which occurred as close as 5 km (3 miles) to the line — but the shooting down of a Jaguar by ground fire on 25 January 1984 caused a change to the rules of engagement. Thereafter, as French troops moved forward to the 16th parallel, they were authorised to fire without warning at any hostile soldiers that approached.[21]

A pattern of French response is therefore emerging, based not on counter-insurgency but upon military presence to protect 'vital ground' to the south of the latest defence line while the politicians search for a long-term solution. Whether they will succeed is open to speculation — a tour of neighbouring states by the French Foreign Minister Claude Cheysson in early 1984 failed to make much headway — but there is no doubt that Chad has all the characteristics of an intractable problem which may yet draw the French into renewed and bitter fighting. Without a policy of counter-insurgency, combining political and military actions in an attempt to persuade the people of Chad to support a central government of mixed-race composition, the chances of solution are remote, yet all the evidence of the last 20 years implies that the French, recalling all too clearly the trauma of the Algerian defeat, will not provide this. The lesson is clear: the adoption of rigid, structured approaches to the complex subtleties of modern insurgency is fraught with dangers which, if realised, can lead to rejection of all aspects of a counter-insurgency doctrine. As *guerre révolutionnaire* undoubtedly contained within it elements of success which could have been applied to Chad, this is indeed unfortunate.

Notes

1. A. Horne, *A Savage War of Peace. Algeria 1954–1962* (Macmillan, London, 1977), pp. 23–8.

2. J. Gottmann, 'Bugeaud, Gallieni, Lyautey: The Development of French Colonial Warfare' in E.M. Earle (ed.), *Makers of Modern Strategy* (Princeton, 1960).

3. M. Charlton and A. Moncrieff, *Many Reasons Why: The American Involvement in Vietnam* (London, Scolar Press, 1978), Chapter 1.

4. Details on the origins of the war may be found in J. Davidson, *Indo-China: Signposts in the Storm* (Longman, Malaysia, 1979), pp. 9–36.

5. Vo Nguyen Giap, *People's War, People's Army* (Frederick A. Praeger, New York, 1962) contains the best overall coverage of Viet Minh origins and organisation.

6. The course of the fighting between 1950 and 1954 is covered in Davidson, *Signposts in the Storm*, pp. 57–83 and E. O'Ballance, *The Indo-China War, 1945–1954* (Faber and Faber, London, 1965), *passim*.

7. The most comprehensive account of the Dien Bien Phu battle is still B. Fall, *Hell in a Very Small Place. The Siege of Dien Bien Phu* (Pall Mall, London, 1966).

8. Ibid., p. 8.

9. For an overview of French counter-insurgency techniques, see E.R. Holmes, 'War without honour? The French Army and counter-insurgency', *War in Peace* (partwork), Issue 36 (Orbis, London, 1984), pp. 714–7.

10. Trinquier's views were later consolidated in *Modern Warfare: A French View of the Counterinsurgency* (Frederick A. Praeger, New York, 1963); Lacheroy's key article was 'La guerre révolutionnaire', in *La défense nationale* (Paris, 1958)

11. G. Bonnet, *Les guerres insurrectionnelles et révolutionnaires* (Payot, Paris, 1958), quoted in P. Paret, *French Revolutionary Warfare from Indochina to Algeria. The Analysis of a Political and Military Doctrine* (Pall Mall, London, 1964), p. 10. Paret is by far the best source in English on *guerre révolutionnaire*.

12. Quoted in Paret, *French Revolutionary Warfare*, p. 18.

13. For more detailed coverage of all aspects of the Algerian War, see Horne, *A Savage War of Peace* and E. O.'Ballance, *The Algerian Insurrection, 1954–1962* (Faber and Faber, London, 1967).

14. Accounts of the 'Battle of Algiers' may be found in Horne, *A Savage War of Peace*, pp. 183-207 and J. Massu, *La Vrai Bataille d'Alger* (Plon, Paris, 1971), although the latter is, obviously, a partisan source.

15. For details of the military aspects of counter-insurgency in Algeria, see Holmes, 'War without honour?' and A. Horne, 'The French Army and the Algerian War, 1954–62', in R. Haycock (ed.), *Regular Armies and Insurgency* (Croom Helm, London, 1979) pp. 69–83.

16. Quoted in Horne, *A Savage War of Peace*, p. 165.

17. Details of the *force d'intervention* and its early deployments are from A. Gavshon, *Crisis in Africa. Battleground of East and West* (Penguin Books Ltd., Harmondsworth, London, 1981), p. 175.

18. For background on Chad, see P. Hugot, 'Les guerres du Tchad, 1964–1983', *Études*, October 1983, pp. 303–16.

19. Figures form P. Younghusband, 'The Chad War' in A.J. Venter (ed.), *Africa at War* (Old Greenwich, Conn., 1974), p. 63.

20. E. Bergot, *La Coloniale du Rif au Tchad, 1925–1980* (Presses de la Cité, Paris, 1982), pp. 241–3.

21. Recent operations in Chad are covered in *Strategic Survey 1983–1984* (International Institute for Strategic Studies, London, 1984), pp. 104–7. For pictorial coverage, see *TAM*, October and November 1983.

References

A. *Indochina*
Charlton, M. and Moncrieff, A. *Many Reasons Why: The American Involvement in Vietnam* (Scolar Press, London, 1978)
Davidson, J. *Indo-China. Signposts in the Storm* (Longman, Malaysia, 1979)
Ely, P. *L'Indochine dans la tourmente* (Plon, Paris, 1964)
Fall, B. *Street Without Joy. Insurgency in Indochina, 1946–1963* (Stackpole, Harrisburg, 1963)
Fall, B. *Hell in a Very Small Place. The Siege of Dien Bien Phu* (Pall Mall, London, 1966)
Giap, Vo Nguyen. *People's War, People's Army* (Frederick A. Praeger, New York, 1962)
Gottmann, J. 'Bugeaud, Gallieni, Lyautey: The Development of French Colonial Warfare' in Earle, E.M. (ed.), *Makers of Modern Strategy* (Princeton Univ. Press, Princeton, 1960)
Langlais, P. *Dien Bien Phu* (France-Empire, Paris, 1963)
Lartéguy, J. *Les Centurions* (Presses de la Cité, Paris, 1960)
Navarre, H. *Agonie de l'Indochine* (Plon, Paris, 1956)
O'Ballance, E. *The Indo-China War, 1945–1954* (Faber and Faber, London, 1965)
Pimlott, J.L. 'Ho Chi Minh's Triumph' in Thompson, R. (ed.), *War in Peace* (Orbis, London, 1984), pp. 62–80
Roy, J. *La bataille de Dien Bien Phu* (Julliard, Paris, 1963)
Scholl-Latour, P. *Death in the Ricefields* (Orbis, London, 1979)

B. *Guerre Révolutionnaire*
Bigeard, M. *Contre guérilla* (Plon, Algiers, 1957)
Bonnet, G. *Les guerres insurrectionnelles et révolutionnaires* (Payot, Paris, 1958)
Hogard, J. 'Guerre révolutionnaire et pacification', *Revue Militaire d'Information*, no. 280 (January 1957)
Holmes, E.R. 'War without honour? The French Army and counter-insurgency', *War in Peace* (partwork), Issue 36 (Orbis, London, 1984), pp. 714–7
Lacheroy, C. 'La guerre révolutionnaire' in *La défense nationale* (Paris, 1958)
Paret, P. *French Revolutionary Warfare from Indochina to Algeria. The Analysis of a Political and Military Doctrine* (Pall Mall, London, 1964)
Trinquier, R. *Modern Warfare: A French View of the Counterinsurgency* (Frederick A. Praeger, New York, 1963)

C. *Algeria*
Behr, E. *The Algerian Problem* (Hodder and Stoughton, London, 1961)
Heggoy, A.A. *Insurgency and Counterinsurgency in Algeria* (Indiana University Press, Bloomington, 1972)
Horne, A. *A Savage War of Peace. Algeria 1954–1962* (Macmillan, London, 1977)
Horne, A. 'The French Army and the Algerian War, 1954–62' in Haycock, R. (ed.) *Regular Armies and Insurgency* (Croom Helm, London, 1979), pp. 69–83
Humbaracci, A. *Algeria. A Revolution that Failed* (Pall Mall, London, 1966)
Massu, J. *La Vrai Bataille d'Alger* (Plon, Paris, 1971)

O'Ballance, E. *The Algerian Insurrection, 1954–1962* (Faber and Faber, London, 1967)

Pimlott, J.L. 'The Algerian Revolution' in Thompson, R. (ed.), *War in Peace* (Orbis, London, 1984), pp. 121–35

Roy, J. *The War in Algeria* (Grove Press, New York, 1975), rept. Hutchinson, London, 1975

Salan, R. *Mémoires: Fin d'un Empire*, vol. 3. 'Algérie française' (Paris, 1972) and vol. 4, 'Algérie, de Gaulle et Moi' (Paris, 1974)

Servan-Schreiber, J.-J. *Lieutenant en Algérie* (Paris, 1957)

Talbott, J. *The War without a Name: France in Algeria, 1954–1962* (Faber and Faber, London, 1981)

D. *Chad*

Bergot, E. *La Coloniale du Rif au Tchad, 1925–1980* (Presses de la Cité, Paris, 1982)

Gavshon, A. *Crisis in Africa: Battleground of East and West* (Harmondsworth, London, 1981), especially pp. 169-83

Hugot, P. 'Les guerres du Tchad, 1964–1983', *Études*, October 1983, pp. 303–16

Strategic Survey, 1983–1984 (International Institute for Strategic Studies, London, 1984), pp. 104–7

TAM, October 1983 ('Operation Manta. Les Soldats de la Paix'), pp. 8–11

TAM, November 1983 ('Biltine: Avec les Soldats de Desert'), pp. 37–43

Younghusband, P. 'The Chad War' in Venter, A.J. (ed.), *Africa at War* (Old Greenwich, Conn., 1974), pp. 60–72

3 THE AMERICAN ARMY: THE VIETNAM WAR, 1965–1973

Peter M. Dunn*

When analysing America's Vietnam War, one is reminded of the fable of the six blind men and the elephant: all were partially right in their interpretations of what they had on their hands and all were wrong in that none was able to describe an animal he had never seen. In the United States, the Vietnam conflict has been variously described as a revolutionary/protracted war, a counter-insurgency, a conventional war or a limited war: all contain an element of truth, but none on its own provides a complete picture. What is apparent is that it was never a straightforward process of insurgents versus security forces, and this makes any study of American counter-insurgency techniques extremely difficult.

Part of the problem is that, in South Vietnam, the indigenous insurgents — the Viet Cong (VC) — functioned both as revolutionaries and as partisans, being purely neither one nor the other. As a minority revolutionary group they fulfilled the classic role of the revolutionary guerrilla, following the pattern laid down by Mao Tse-tung in China, but they also acted as partisans for a large foreign regular army (that of communist North Vietnam), forming battalions, regiments and even divisions when the need arose. This forced the US effort in Vietnam to be diversified, dealing simultaneously with the VC insurgency in the South, the substantial conventional military threat from the North and the elaborate supply line (the Ho Chi Minh Trail) which linked and sustained the enemy fronts. At no time could the concentration of force so essential to success be achieved.

The result was a series of separate campaigns, fought all at the same time by different elements of the security forces. The Army, for example, conducted a conventional war against

---

* The views expressed in this article are those of the author and do not reflect the official policy or position of the Department of Defense or the US Government.

Map 3.1: Vietnam

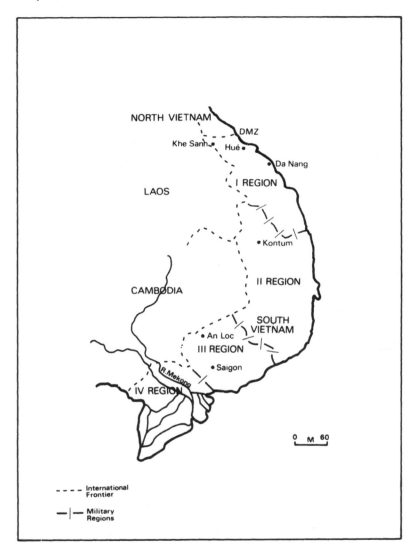

conventional enemy units, a war against guerrilla forces, a war to win hearts and minds and a war to bolster up South Vietnam before the enemy destroyed it. The Marines, who strongly disagreed with the Army's approach, had a separate campaign going on in the northern quarter of South Vietnam, involving the same factors but a different philosophy, while in the air several wars were going on at once — over North Vietnam, Laos, Cambodia and the South — all under separate (and invariably parochial) command structures, involving elements of the Air Force, Navy, Marines and even the Army. The US Army itself could not therefore conduct a counter-insurgency war as it is generally understood and the only units that came close to doing so — the Special Forces — tended to operate in a strategic vacuum, often in remote, sparsely-populated border regions which were physically removed from the core of the counter-insurgency problem. Interestingly, the Marines, despite their primary mission as a mobile assault force, came close to understanding the response to insurgency in Vietnam, but were never integrated into the wider, predominantly Army, operations. Similarly, although a plethora of non-military agencies — the Central Intelligence Agency (CIA), the Agency for International Development (AID), the State Department and many others — appeared in the field, their actions were rarely tied to those of the military. In short, the absence of close civil-military co-ordination, coupled to the diverse nature of the threat, undermined the development of a workable counter-insurgency strategy.

Not that the Americans were unaware of the nature of counter-insurgency. In the late 1940s, for example, a liaison team — the Van Fleet mission — had been sent to Greece to assist the government forces against communist attack and although its primary mission was a logistical one, it did gain a close view of the strategies and tactics employed by both sides. Of even more importance was the experience gained during the Korean War (1950–3) against North Korean guerrillas operating in the mainly mountainous terrain of southwest South Korea — guerrillas who bore a number of resemblances to the VC of the 1960s — for the lessons drawn from these operations were precise and relevant. In 1955 it was noted that, when presented with a guerrilla threat, the security forces had to bear the following points in mind:

1. The nature, objectives, tactics and vulnerabilities of the enemy must be recognised.
2. A broad policy, combining military action conducted by adequate specially-trained forces under dynamic leadership employing political, economic and psychological measures designed to gain the support of the civilian population, and isolate and destroy the guerrillas, must be adopted.[1]

Furthermore, it was emphasised that military operations alone were not enough to defeat the insurgents. It was understood that an inter-agency, combined approach at the governmental level had been necessary in Korea to tackle all aspects of the civil-military problem and that this had eventually succeeded in denying the popular support so essential to the guerrillas. Finally, success would not have been possible without 'a thorough knowledge of the needs, customs and beliefs of the people'.

Thus, although a direct comparison between Korea and Vietnam is impossible — the VC, for example, had a political infrastructure which was far more sophisticated and difficult to deal with than their Korean counterparts — it is apparent that the basic principles of counter-insurgency had been appreciated. Indeed, there were those in the American Services who recognised that the sprouting communist-led insurgencies of the 1950s and early 1960s could be — and had been — defeated and understood the common themes linking the various campaigns; there were common reasons why they arose, why they succeeded and why they failed. Years before the Indochinese communists forced Americans in general to study the fundamentals of rural-based revolutionary war as promulgated by Mao (and fine-tuned by the Vietnamese communists), at least some US officers — notably Colonel (later Major-General) Edward Lansdale, who had gained personal experience in the early 1950s in the Philippines, aiding Defence Minister (later President) Ramon Magsaysay against the communist Hukbalahaps — understood the 'fish and water' Maoist analogy and the principles for defeating communist insurgency. Unfortunately they were in a minority and could not be heard amid the crowd of big-weapon and big-bang generals, few of whom understood that the war of the Space Age was in fact the war of antiquity — irregular, guerrilla war.

As American involvement in Vietnam developed in the

aftermath of the French defeat at Dien Bien Phu (May 1954), therefore, the nature of the communist threat and the scope of possible response to it was totally ignored. No attempt was made, for example, to seek the advice of the French nor to absorb their bitterly-learned lessons, while in Washington the higher echelons of the armed forces refused to listen to those who understood the need of counter-insurgency. By 1964, as the political chaos of South Vietnam and the development of threats from the VC and the North emerged, the Americans found themselves drawn into a maelstrom of conflict for which they were not prepared. This was reflected in their approach to the problem in political, military and socio-economic terms, and although some valid ideas were to emerge during the period of force commitment to Vietnam (1965–73), the overall campaign was muddled and poorly co-ordinated. The American experience of counter-insurgency was one of deep frustration as a host of different military commands and civilian agencies sought in vain to create a pattern of response which would answer the problem of a war of the most complex mutation. The ramifications of the experience are still apparent today.

Perhaps no topic generates more heated discussion among American military men than that of command and control. From the 'committee' approach to military operations and control in Vietnam to the special study groups and *ad hoc* committees in Washington, the American approach to this aspect of the war was marked by cluttered chains of command and the inability of institutions and Services to relinquish power. The Washington bureaucrats thought that because they could communicate instantly with far-off headquarters, they could make sound tactical decisions half a world away from the battle zone; the Marines and Army disagreed on strategy; the Air Force, Marines, Navy and Army had differing views on the use of air power and the other Services would not agree to the Air Force controlling the air war in Vietnam; even within the Air Force, generals would not relinquish control of their airplanes to the Air Force commander in Vietnam.

At the same time, one of the most notable aspects of this war was its 'amateurisation' by civilians in Washington. These civilians, most without military experience or training, dabbled to an unheard-of degree in every aspect of the war, and every government agency seemed to be involved. Sir Robert

Thompson, the British expert on counter-insurgency, once remarked that despite all his visits to Washington, he could never be quite sure who was really in charge of the war — the White House, Pentagon, Congress, the State Department or others; all had bigger or smaller bits of it, but no one seemed to be really in charge. As one monumental study put it:

> The war in Vietnam was unique in many respects, not least of which were the multiple and sometimes unorthodox command and control arrangements. At the peak of the US involvement in late 1968, there were over 1.6 million South Vietnamese, US and other Free World military personnel concentrated in the 600,000 square miles of RVN; no single person or agency was in overall charge of them.[2]

The situation was no better in Vietnam itself, for although the US Ambassador in Saigon was nominally the senior US representative in the country, in effect he left the conduct of the war to the Military Assistance Command, Vietnam (MACV). In fact, Ambassador Maxwell Taylor, a retired four-star Army general, initially disagreed with the strategy adopted by General William Westmoreland (MACV Commander, 1964–8), but did not intervene to have this strategy altered. However, although Westmoreland had a fairly free hand within South Vietnam, he could not use his regular forces outside the country, and the State Department in Washington took to interfering in purely military affairs,[3] exploiting the fact that there were fundamental differences of opinion as to how MACV should be structured and how it should be treated. The Joint Chiefs of Staff (JCS) wanted MACV to be a unified command reporting to them, while the Commander-in-Chief, Pacific (CINCPAC — an admiral) wanted it under him. The State Department leaned to the CINCPAC view, but even so, wanted their ambassador to be the senior authority in South Vietnam. There were other proposals for command arrangements which found disfavour in other quarters, and it took several months of debate before MACV was finally declared a unified command under CINCPAC. This meant that Westmoreland was responsible to two immediate superiors — CINCPAC and the Commander-in-Chief, US Army, Pacific — and although MACV was under CINCPAC, communications in fact went directly to the JCS or higher, with messages going simultaneously to CINCPAC and vice versa.

Nor did the confusion end there. When the Marines landed at Danang in 1965, their commander, a brigadier-general, was designated the naval component commander under MACV; however, later that year a Navy rear-admiral was appointed Chief, Naval Advisory Group, MACV, which meant that until February 1966, Westmoreland in effect had two naval component commanders under him. Similarly, bombing operations in the theatre came under several separate controls. CINCPAC controlled the bombing of North Vietnam, but certain tactical targets had to be approved by the White House. The MACV Commander controlled the bombing in the South and in Laos, but carrier aircraft were under 7th Fleet and CINCPAC, and B-52s were under the Commander-in-Chief, Strategic Air Command, headquartered in Nebraska. In the northern sector of South Vietnam the Marines even controlled their own air wing, larger than an Air Force wing. Thus the Air Force Commander in Vietnam controlled but a fraction of the aircraft bombing Indochina.

At another level, the proposal to establish a unified command among the Allies arose repeatedly, but Westmoreland opposed the formation of a joint US-Vietnamese Command on the Korean War model, so that at no time did he have operational control of Vietnamese forces, nor indeed the South Korean divisions or the Philippine Civic Action Group, all of which remained under autonomous command structures. Some coordination was achieved in theory between the Americans and South Vietnamese through the existence of 'advisers' attached to ARVN (Army of the Republic of Vietnam) unit, but the officers involved were of erratic quality, some being young lieutenants straight from basic training in the United States.

Nine advisory groups reported to MACV; the Air Force and naval advisers reported through their Service component commanders who controlled their efforts, but the Army advisers did not report to the Army component commander when the latter's position was eventually created — they reported directly to the MACV Commander. By the same token, when the Civil Operations and Rural Development Support (CORDS) organisation was formed in 1967, it too was placed under MACV, with the CORDS civilian head becoming Westmoreland's deputy for pacification.

Finally, many civilian agencies — the CIA, AID, the State Department and more — while nominally under the US Ambassador or MACV, usually had direct and independent communications to their own higher headquarters in Washington. For a while Westmoreland's deputy for CORDS, Robert Komer, was sending to Washington drafts of directives he wanted Washington to send to the MACV Commander — until Westmoreland found out about it and stopped it.

The problem of clear lines of command and control plagued the American effort to the end, with senior military officers, the White House Staff and others, all constantly bypassing existing channels to communicate directly to the field in Vietnam: one Special Forces officer at a critical moment in a remote camp in the Central Highlands reported receiving a call from a White House staff officer. These queries, perhaps thousands per year from various sources, warranted replies and were an enormous drain on the various headquarters in Vietnam. They also represented a degree of uncoordinated interference in tactical operations — ranging from restrictions on the targets, bomb-loads and routes of aircraft over the North down to the unnecessarily complex 'Rules of Engagement' for troops on the ground — which inevitably arose from a system bereft of close political centralisation within a framework of counter-insurgency.

This inefficiency could have been avoided, for it appears that the seriousness of this lack of overall command and control was appreciated. Early in the war, when the chronic deficiencies of the 'American way of war' became apparent, President Lyndon Johnson and Defense Secretary Robert McNamara attempted to place the war theatre under a single overall politico-military authority. General Maxwell Taylor was given the chance to be, in effect, a proconsul on South Vietnam, but instead of seizing the opportunity, Taylor assured American commanders-in-chief that he would not use these powers.[4] His refusal was predictable, but was undoubtedly an error. As Taylor himself later admitted, 'one of the facts of life about Vietnam was that it was never difficult to decide what should be done but it was almost impossible to get it done'.[5]

Such confusion at the top inevitably affected the conduct of the military campaign, but there was more to it than that. The rival armies in Vietnam had radically different roots. It has been said (probably correctly) that the US Army's instincts went back to the American Civil War of the 1860s, while the Vietnamese communists drew heavily from Mao's theories and practices of protracted war. So while the North Vietnamese campaign of insurgency in South Vietnam was built on continuously refining the practice of protracted, revolutionary war, the US Army was institutionally unable to adapt to this kind of warfare. By its more conventional response, its strategy of attrition and the unceasing quest for the big set-piece battle, the Army became, in effect, a large French Expeditionary Corps — and met the same frustrations.

US Army doctrine paid scant attention to unconventional warfare. With little interest or recent experience in counter-insurgency on a large scale — and few recognisable pay-offs in annual budgetary allocations or personal career enhancement — the evolving US Army strategy was perhaps inevitable. Although other metaphors have been used, the Army was going to use a sledgehammer to crush a fly, while the practice of unconventional war was left largely to the Special Forces.

Of the US regular forces which went into South Vietnam in strength in 1965, only the Marines developed a strategy of mounting what came close to a proper counter-insurgency campaign. This was born of necessity when the Marines realised that they could not defend Danang without getting involved with the people in the countryside. The Army had no real ideas on how to fight this war, and looked only to finding enemy regular 'main force' units — a generally fruitless task as it turned out. Their 'search and destroy' operations by large units were costly to mount, spectacular to see and generally unproductive.[6] The Army's strategy was controversial, even within the Army. Many, including the Chief of Staff, General Harold Johnson, and former generals James Gavin and Maxwell Taylor, would have preferred to see US troops concentrate on securing the main centres of population, which largely lay along the coasts (more or less an 'enclave' strategy), but Westmoreland preferred a more active role and adopted a strategy aimed at killing so many of the enemy that the communists would be unable to continue fighting — a strategy of attrition.

The Army established itself in major bases throughout the country, from coastal bases to the Central Highland strongholds of Pleiku and Ankhe (where the 1st Air Cavalry Divsion was based). In this 'non-linear' war, there were no extended field campaigns as known in previous wars. Operations were mounted from these bases and were generally measured in terms of hours or days rather than in weeks or years. These US bases were like the cavalry forts of the American West in the late nineteenth century, from which troops of cavalry sallied forth periodically to search for the Indians. In Vietnam the soldiers sallied forth in helicopters or amoured personnel carriers to search for the enemy; in fact, hostile territory came to be called 'Indian country'.

The firepower brought to bear on the enemy — when he could be found — was almost incomprehensible in its enormity and its disparity when measured against the firepower available to the enemy. The whole range of modern technology and fire support was available to American commanders. US and Korean forces made lavish use of this firepower, and target areas were generally pounded ('prepped') by intensive artillery and air bombardment prior to assault by infantry. This avoided friendly casualties but sacrificed the principle of surprise. At night, harassing artillery fires were usually laid on preselected routes and areas throughout the country, which caused discomfort to both the enemy and rural population alike.

US troops rarely fought at night. Although individual units occasionally laid night ambushes, in general the night belonged to the communists — as it always had in Indochina. This kind of strategy left the civil population in an unenviable position; since the Allied troops maintained themselves in their bases at night and when not engaged in operations, the communists had a fairly free hand between sunset and sunrise. It was thus difficult to win the allegiance of the peasants when they could be protected for only half a day, even if the social and historical bases of the revolutionary war had been understood. Although American soldiers felt superior in every way to the Vietnamese, in fact they were far less politically sophisticated than their communist opponents.

The concept of air assault, developed by the Army in the 1950s, matured — and on occasion over-ripened — in Vietnam with the formation of an 'Air Cavalry' division and the use of air assault tactics throughout Vietnam. Air assault, to the Army, was a marriage of firepower and manoeuvre,[7] and the development of the huge CH-54 'Flying Crane' helicopter allowed the heavier 155 mm artillery pieces to move with the air assault units.

Helicopter gunships, modified to fire rockets, were added to the air cavalry units and were known as 'aerial artillery'.

American infantry rode when not flying, for most Army divisions were mechanised. While the advantages in speed, protection and mobility were obvious, terrain such as the hills, jungles and paddies of Vietnam were not ideal for the employment of armoured vehicles. The disadvantage was that 'mechanised infantry units . . . equate ability to move with vehicle maintenance'.[8]

In addition to the masses of helicopters and various vehicles available, Air Force transports — Caribous, C-123s and C-130s — capable of landing on short, unprepared surfaces could haul men and material around Vietnam, and a large fleet of jet transports, supplemented by chartered civil aircraft, plus the merchant ships, moved supplies and troops quickly and efficiently to Vietnam. It proved again the American penchant for logistical, if not strategic or tactical, brilliance.

A variety of special-purpose units sprang into existence; these included the Mobile Riverine Force (composed of naval and army units operating along the vast Mekong Delta water systems), Air Force Air Commando squadrons and gunships, and more; but, like the more conventional units, they addressed primarily the military dimension of the revolutionary war, and at the war's end they were for the most part disbanded as rapidly as they had been formed.

But the strategy of looking for the enemy failed, and the enemy was found generally when he wanted to be found. One reason for this was that the VC military forces and Communist Party apparatus operated in their home provinces — fish in home waters — while the South Vietnamese and Americans were generally strangers in their areas of responsibility. In 1967–8, for example, less than one per cent of the nearly two million reported small-unit operations resulted in contact with the enemy. Despite the fact that the number of US 'battalion days of operations' increased dramatically in 1968, the percentage of contacts with the enemy decreased to a 'remarkable degree' according to the CIA. Some in the Army argue that the low percentage of contacts was because the enemy had been severely mauled in the Tet offensive of early 1968; this was partly true, but the evidence indicates that the percentage of Allied-initiated contacts remained low throughout the war.

The enemy controlled the casualty rates for both sides; thus not only was the US attrition strategy relatively unproductive, it was the enemy who actually controlled the attrition. In 1969 the American strategy began to change after the elevation of General Westmoreland to Army Chief of Staff and his return to Washington. The United States was seeking to disengage from the war and General Creighton Abrams, Westmoreland's successor as MACV Commander, was charged with decreasing the active American combat role and speeding up 'Vietnamization' in preparation for an eventual US withdrawal. The defeat of the communist attack during Tet in 1968 had been an empty military success for the Allies, since the communists achieved the victory that counted — the psychological and political victory.

Like the French before them, the United States kept it until very late in the day to start serious Vietnamization — to strengthen and prepare the South Vietnamese armed forces to assume the full burden of self-defence. Dr Robert Sansom noted that 'in 1969 I was visiting MACV, trying to find out how much of the staff echelon were working in Vietnamese tables of organisation, equipment and the rest of it. I was amazed to find out how recently it had begun, how little had gone into this sort of thing.'[9] Sansom further noted that there had been a Vietnamese Army since 1952, but that little thought had gone into it and the US 'had done it so badly . . . in the 1960–65 period'. Major-General George Keegan, former Chief of US Air Force Intelligence, added: '. . . we trained an army and air force. Wrong equipment, wrong tactics, maybe, wrong doctrines. . .'[10]

At the height of American combat involvement in Vietnam there were sharp differences between the CIA, Defense Intelligence Agency (DIA) and the State Department on the one side, and MACV on the other, in establishing the VC and North Vietnamese Army (NVA) Order of Battle. The MACV figures were significantly lower than those derived by the national intelligence agencies; this was because military intelligence (at MACV), despite strong internal disagreement by lower-ranking officers, discounted the effect of the communist irregular, paramilitary forces — Self-Defense Forces, Assault Youth, etc. — and refused to include them as a threat. The CIA disagreed with MACV and wrote that while these forces 'are not of the same military significance as combat and support troops or guerrillas . . . nevertheless, they do perform important military

support functions, inflict and receive casualties, and are meaning-
ful elements of the enemy's organized resistance'.[11]

One problem never satisfactorily resolved was the transmission
of 'national level' intelligence to the field. Intelligence from
sophisticated technical systems was often considered too sensitive
to be sent to tactical units in the field and the information often
sat, effectively useless, in an office safe in headquarters. Friendly
losses occasionally occurred as a result of the witholding of data
of tactical interest to the operating forces.[12]

While the lack of a unified command structure remained a
problem, there was one area where a major joint US/South
Vietnamese endeavour was established and worked well; this was
in the field of intelligence (including counter-intelligence). At the
suggestion of then Brigadier-General J.A. McChristian, the
MACV Assistant Chief of Staff, Intelligence (J-2), '. . . we
would establish centers throughout the country for interrogation
of prisoners and *Hoi Chanhs*, "ralliers to South Vietnam", and
for exploitation of captured documents and material as well as
centers where all information would be sent for collection,
analysis, evaluation and processing into intelligence in support of
US and South Vietnamese Forces'.[13] Like the Marines' Combined Action Program, this worked relatively well, with each side
compensating for the other's weakness. The Americans would
never understand the land like the South Vietnamese, while the
US could draw on sophisticated 'national technical systems' —
satellites, U-2 reconnaissance planes, signals intelligence
(SIGINT), etc. to look at the larger picture.

Communications and information security, however, were
generally less than satisfactory, and the US forces frequently paid
scant attention to these areas. General McChristian later wrote,
'. . . it is doubtful that the average US officer or enlisted man
ever appreciated the extent of the Communist collection effort
even though the Counterintelligence Division placed maximum
emphasis on educating them to the security hazards confronting
the command daily'.[14] Further, the rather simple, unsophisticated
character of the enemy disguised his complex, highly efficient
intelligence system:[15] all in all, collecting intelligence against US
forces (in the United States as well as in Vietnam) should not
have been a particularly difficult problem for the enemy.[16]

The intelligence support of the first large (multi-division) Army
operation — 'Cedar Falls' (January 1967) — illustrates both a

text-book case of what intelligence should be and the US Army strategy for the war. Cedar Falls was unique also in that it was conceived by an intelligence officer,[17] its objective being 'the capture or destruction of Headquarters, Viet Cong Military Region IV and its base camps and supply bases as well as the 272nd Viet Cong Regiment'. Since the villagers were reported to be supporters and helpers of the VC (willingly or otherwise), the village of Ben Suc was to be destroyed, thousands of people moved and the whole triangular area was to become a 'free fire zone'. An intensive intelligence collection effort, using all sources, had pointed out the danger to Saigon posed by these communist forces in the 'Iron Triangle' so near the capital.

The Iron Triangle lay 56 km (35 miles) north-northwest of Saigon; it was bounded by the Saigon and Thi Tinh Rivers and by the Than Dien forest reserve in the north. The Triangle's corners were anchored by the villages of Ben Suc and Phu Hoa Dong on a northwest-southeast axis along the Saigon River, and by Ben Cat, east of Ben Suc and north of Phu Hoa Dong, along the Thi Tinh River. The operation was to have two phases: the first (lasting from 5 to 8 January 1967) involved the surrounding of the operational area, while the second (from 8 January for about two or three weeks) was the assault and clearance of the Iron Triangle. Some 15–20,000 Allied troops were involved.

The US commanders tried to disguise the operation by conducting masking operations in the region, and by 8 January the blocking forces (25th Infantry Division elements and 196th Light Infantry Brigade) were in place along a 24 km (15 mile) area south of the Saigon River — from the Boi Loi and Ho Bo woods across the river from Ben Suc southwest to the Filhol Plantation and Phu Hoa Dong village. These forces were the 'anvil'; the 'hammer' consisted of the 1st Infantry Division air-assaulting in an arc from the north and northwest to Ben Suc and the Than Dien forest reserve, plus a westward drive across the top of the Triangle by the 11th Armored Cavalry Regiment and 173rd Airborne Brigade.

Although pains were taken to achieve surprise, to include the witholding of information to the ARVN until the day before the operation, the air assault into Ben Suc occurred one day before the armoured hammer was to strike across the Triangle. Further, 20 hours before the assault on Ben Suc, the 1st Division public affairs officer had announced the plan to the press, and the day

Map 3.2: The 'Iron Triangle'

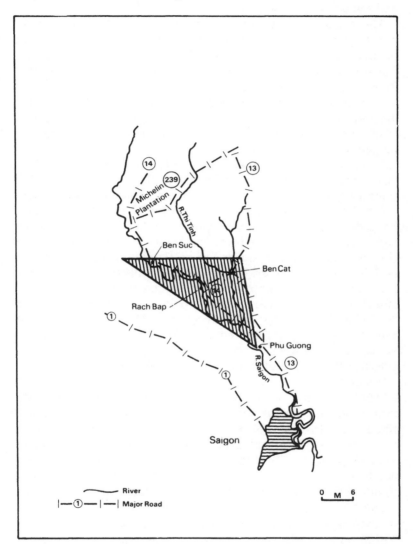

before the hammer swung, the customary air and artillery strikes had signalled an attack was imminent.

The commanding generals claimed complete tactical surprise, but it appears that only the villagers were caught unawares. The several VC main force battalions, following Mao's principle, had vanished as the enemy advanced, and only desultory resistance from residual communist local forces was encountered. Vast tunnel complexes were destroyed, valuable intelligence gained in the form of captured documents, and large amounts of enemy food and supplies were seized, but the enemy battalions had slipped away to fight another day.

Ben Suc was evacuated and levelled and the Triangle searched; the operation ended on 26 January. The tally showed 750 confirmed 'enemy' dead, 280 prisoners captured, 540 'ralliers' and 512 suspects detained — a total of 2082 people — yet only 590 individual weapons were captured, indicating to some that a high proportion of non-fighters had been killed or detained. The US commanders called Cedar Falls a resounding success and the 1st Division commander, Major-General William DePuy (himself a key proponent of the American attrition strategy in Vietnam) said that it was a 'decisive turning point . . . a blow from which the VC in this area may never recover'.[18] The fact that within 48 hours VC were again reported in the Iron Triangle, following the Maoist rule that when the enemy retreats the guerrillas advance, indicates the failure of the American approach.

It was perhaps ominous that the very beginnings of the war in Vietnam were marked by disagreement on how to approach the socio-economic dimension (pacification and winning hearts and minds). In retrospect, this was to be expected since there were no contemporary American models for this extremely complex evolving problem. The Malayan insurgency, managed with great skill by the British (1948–60), was a relatively simpler and smaller affair, with the numbers of communist terrorists being small, ethnically distinct and often bereft of sanctuary. The Hukbalahaps in the Philippines (1946–54), also defeated by the skilful and imaginative President Magsaysay, were likewise confined to islands. In Vietnam the dimensions of the insurgency were immensely larger and more difficult to define, and all things were possible, including a massive invasion by North Vietnam's large, tough and disciplined regular army. This is why the British and Filipino experiences were only of limited value in Vietnam.

In the 1960s Robert Thompson, of the British Advisory Mission, Saigon, developed a counter-insurgency plan with which the American mission disagreed. The Thompson paper reflected the slower, more patient approach so successful in Malaya, and one which emphasised the primacy of civil authorities, particularly the police forces, in the struggle. The Americans wanted to move more rapidly against the guerrilla military forces. Thompson's idea was to win the allegiance of the rural peasant population, for they in effect were the battleground, while the Americans wanted to seek out and destroy communist military forces.

Thompson's idea was to stack the deck in one's favour by starting in an important area not yet heavily penetrated by the communists (the Mekong Delta), while the Americans wanted to move against a heavily infiltrated area near Saigon. President Ngo Dinh Diem of South Vietnam liked Thompson's idea, but wanted to implement the general plan somewhere else. All ideas involved the 'strategic hamlet' concept, although in Thompson's case these were but a means to an end.

A fundamental disagreement, which persisted to the end, lay in the role of the regular military forces. In the British view, these forces would go in and clear the area of insurgents, then keep them off balance while the area was pacified by government cadres — the 'clear and hold' concept. The Americans, as they did later with their own units, preferred a 'search and destroy' role for the troops — but, as was found in a few years, it was extremely difficult to find the enemy by using large-unit sweeps, and much time, effort and money was to be squandered on these tactics by US forces.

The debate in the early 1960s was reflected in the Kennedy Administration until a decision was made — based largely on a paper by Roger Hilsman — to adopt the sort of approach favoured by the British team. Resources were allocated to a plan, but the South Vietnamese scattered their energies in difficult provinces and the plan failed. Once again, the peasants had not been convinced that these changes were good for them (or were permanent) and the strategy changed with the accession of the Johnson Administration.

But these debates and tactics — particularly concerning the resettlement of populations — had preceded the arrival of the Thompson team in 1960. In the late 1940s and early 1950s the

French had also resettled civilians in the course of their pacification programme, and the South Vietnamese under Diem continued these efforts. By 1959 'agrovilles' of 300 to 500 families were grouped around 'Rural Community Development Centers', and by the early 1960s the agrovilles had become 'strategic hamlets' in the 'Strategic Hamlet Program' (SHP) — which by 1963 had failed. The peasants resented being removed from their ancestral lands. Further, without the bedrock of physical security, no such scheme could work.

Recognising this, the South Vietnamese government armed the 59-man cadres who were the spearhead for the Revolutionary Development (RD) programme (1965–7), and dozens of battalions from the ARVN were assigned to help in clearing and holding areas targeted for pacification.[19] The CORDS Programme was designed to pull together the frequently disjointed efforts at pacification, of which construction, medical and educational assistance and physical security were all parts.

The rural paramilitary forces had earlier begun to receive more attention. In 1964 the old Civil Guard had been renamed the 'Regional Forces' (RF), to operate at the province level in company strength, and the *Dan Ve* (Self-Defence Corps) was renamed the 'Popular Forces' (PF), to be used in platoon strength at the village and hamlet level. These forces were separate from the other main paramilitary group, the Civilian Irregular Defense Group (CIDG), which was sponsored by the Special Forces for missions which eventually grew beyond local self defence.

In 1967 new emphasis was placed on pacification with the establishment of CORDS under Robert Komer (who was given the rank of ambassador and named General Westmoreland's deputy for pacification). So in addition to the several land, sea and air wars being prosecuted in Vietnam, the United States now had two ambassadors in the country; something which did nothing to ease the already-chronic problems of diversified command.

In a defence environment dominated (some said plagued) by 'systems analysts', demands soon arose for some sort of score cards with which to measure the course of the game, and the Hamlet Evaluation System (HES) became to pacification what the notorious body counts were to the Army — not very valid, and often misleading, indicators of 'progress'. The HES began with a required report by the American district adviser; these ratings, by over 250 such advisers, were necessarily subjective and often resulted in erroneous conclusions being drawn by Washington officials eventually given to grasping at straws. Due to the questionnaire's matrix, hamlets could be rated as overall 'satisfactory', when in fact they were not secure. In 1970, in an effort to get a more accurate measurement, the rating system was made more objective and sophisticated with higher echelons forbidden to alter the advisers' reports, but the results were generally the same and the reporting system itself involved a massive effort.

The shock of the communist Tet offensive of 1968, in which the RF and PF were heavily involved, resulted in an increase in these forces to 100,000 men, and by 1973 (the effective culmination of Vietnamisation and the withdrawal of US forces from Vietnam), their numbers had grown to 500,000 men in 1600 RF companies and 7000 PF platoons.[20] Tet brought with it the realisation that the whole country should be armed, that the rural population must be mobilised for self-protection. Thus the moribund People's Self-Defense Forces (PSDF) were regenerated and eventually reached an enrolment of four million men, about a sixth of whom were armed.

Another feature of the pacification programme was the *Chieu Hoi* ('Open Arms') programme, which encouraged enemy soldiers and supporters to defect, or 'rally', to the South Vietnamese. This programme drew in significant numbers of defectors (200,000 since 1963), but of interest was the fact that most of these came from primarily indigenous communist forces in the southern part of the country; the North Vietnamese forces, established in the northern areas, kept a tighter hold on their supporters.

There were other notable aspects of pacification — the political propaganda police work, rural reconstruction and land reform — all successful to varying degrees. The key point to all this, however, was that on the US side these programmes were envisaged, planned and executed largely by civilians; the Army, as in the area of strategy (and occasionally tactics) had by its own deficiencies relinquished responsibility for half the war to civilians.

One of the most controversial aspects of the CORDS organisation was the 'Phoenix' (*Phuong Hoang*) programme, a

strategy for eliminating the VC infrastructure (VCI). Since the
VC had enjoyed success in eliminating thousands of Vietnamese
— village chiefs, teachers and policemen — who were parts of
the government's infrastructure, it stood to reason that a similar
campaign against the communists would produce similar damage
to their infrastructure. While Phoenix was an American idea, it
was run largely by the Vietnamese, and in 1969 the CIA turned
over its advisory role to CORDS. Although thousands of
communist agents and workers were captured or killed under
Phoenix, National Security Study Memorandum (NSSM) 1 noted
that most of the VCI eliminated were lower echelon cadres — the
more important leaders were generally untouched.[21]

Phoenix had its problems: many of those captured were not
convicted by the Vietnamese courts. Despite its controversial
nature, it was not primarily an assassination programme
(although some assassinations were carried out), and many of the
VCI killed were shot while resisting arrest or during fights with
the security forces.[22] The programme of assassination carried out
against the government by the communists was more brutal and
effective, but less publicised by Western newsmen.

At the height of the Phoenix campaign about 30,000 agents
were employed by the CIA and South Vietnamese government;
they, and other Allied police and intelligence agents, were
abandoned by the CIA on the fall of the South in 1975.[23] In
retrospect, the CORDS programme was relatively successful in
Vietnam. Like so much else, it broke down in Washington.[24]

The second major group of paramilitary forces — not
technically part of the pacification programme — was the CIDG.
It was President John F. Kennedy, rather than the US Army,
who took the Soviet concept of 'wars of national liberation'
seriously. Despite the fact that these limited wars of irregular
dimensions have been the most common conflicts for decades,
the President was never able to get the Army seriously to
consider this 'new' type of warfare.[25] It was as a result of
Kennedy's acute interest in unconventional warfare that the
Special Forces were given new life.

Like the President, the civilians in the CIA understood that a
new approach would be needed in Vietnam, and at their request
the Special Forces began operations in Indochina in remote areas
and among peoples with whom the South Vietnamese government had had little success or interest — the *montagnards*

(mountain people) who, over the centuries, had been gradually pushed by the expanding lowland Vietnamese population into the less habitable and less accessible regions of Annam and Cochinchina.

The idea of the CIDG was conceived in 1961, and in early 1962 a Special Forces detachment on temporary duty from Okinawa began work among the largest *montagnard* tribe, the Rhade. When the Special Force team, under Captain Ron Shackleton, returned to Okinawa six months later, they left behind '120 Rhade hamlets protected by close to 10,000 village defenders'.[26] As Colonel Robert Rheault (former commander of the 5th Special Forces Group at Nha Trang) put it, the CIDG programme was:

A story of teaching the Vietnamese how to shoot, build a farm, care for the sick, or run agent operations. It was working with the religious and ethnic minorities of Vietnam: the Montagnards, the Cambodians, the Hoa Hao, and the Cao Dai. The program was both praised and reviled by Americans and Vietnamese alike and was on the verge of being destroyed many times — not by the Viet Cong, but by its creators, the American command; and by early 1971 it had disappeared from the scene.[27]

While civilians had conceived and provided the impetus for the CIDG, 'it was, indeed, the military who later converted (or, as some feel, perverted it) and who finally killed it'.[28]

In 1970 the Special Forces were caught up in controversy surrounding their methods, particularly in a highly publicised case involving the killing of a Vietnamese double agent. The Special Forces commander, among others, was jailed and the entire group returned to the US soon after. In 1971 the CIDG forces were taken over by the South Vietnamese government and became part of their paramilitary forces.

Colonel Rheault later wrote, '. . . worst of all . . . after all these years in Vietnam, MACV still did not understand either the CIDG program or revolutionary war'. Perhaps the death of the programme was inevitable, given the shifting emphasis on the CIDG over the years. It had started out as a village self-defence programme, and as the strength of the regular forces grew, its actions gradually took on a more military flavour. Nevertheless, the Special Forces/CIDG teams accomplished a great deal and if they did nothing else, the fact remains that they lived, operated and gathered intelligence in the kind of remote areas often infested with the enemy and untouched by the regulars. And for the systems analyst it was more cost-effective since it cost less, per enemy body, to kill the enemy using the irregulars.

Although arguments still persist as to the effectiveness of this effort, the Special Forces eventually established 80 CIDG camps throughout the country, with 60,000 irregulars under arms. By any standards that was a good dividend on a relatively small investment, and these forces were a factor in the overall equation. The problem, as ever, was that the programme was something of a stepchild, on its own, not fully integrated into the normal operations and interests of a regular army which was largely ignorant of revolutionary guerrilla war. Indeed, given the complicated budgetary system, it could not have existed at all without the special funding of the CIA.[29] A case could be made, however, that the Special Forces may have produced even greater results if deployed in the hamlets and the centres of rural population, where the war for hearts and minds was being fought out.

The one programme where the regular and paramilitary forces were integrated in a pacification strategy was the amalgamation of the US Marines and the Vietnamese PF units in the (Marine) Combined Action Program, which began in 1965, two years before the formation of CORDS. In some Marine Tactical Areas of Responsibility (TAORs) in the northern quarter of South Vietnam, the Vietnamese Army commander relinquished operational local control of selected PF units to the US Marine battalion commanders. The concept was to build an infrastructure like that of the VC, which as it grew would push the VCI out. The missions of these combined forces included 'security, counter-intelligence, obtaining the good will of the people . . . These formed the spokes of the wheel while training was the hub of the entire operation.'[30] The control arrangements were informal and varied from one District Chief to another, and even where operational control of the PF platoons was assumed by the Marines, the District Chief retained administrative responsibility for the PF.

The subsequent 'joint action companies' (later platoons) consisting of a Marine squad integrated into a PF platoon, were generally patterned after similar British companies used earlier in Malaya. Each reinforced the other's strengths and compensated for the other's weakness, and combined with local medical assistance, reconstruction, etc. made an impact in this war of and for the people. The intelligence received by the Allies grew better, the farmers were protected, the communists were denied food and taxes and a relatively low investment paid high dividends. While the Marines retained their belief in pacification, in some TAORs the emphasis by necessity was on the big-unit actions, and the efforts on pacification varied with the Marine division commanders.

General Westmoreland has written that while he appreciated the combined action concept, he did not have sufficient numbers of soldiers to deploy in this manner throughout the country; it was a matter of priorities. Given the high number of support personnel required to sustain the Army's actual fighters — the penalty imposed on what many saw as an unnecessarily sophisticated army — this was probably correct, but does not alter the fact that, with some notable exceptions, the American approach to the all-important socio-economic aspects of counter-insurgency failed. Even in those areas which promised success, the advantage could not be exploited in the absence of centralised co-ordination and control — arguably the key weakness in the American campaign.

When all is said and done, perhaps the most notable feature of America's twelve-year, multi-billion dollar and multi-million man effort in Vietnam is what little impact it has had on strategic thinking in the US Army and Marine Corps. Doctrinal papers of the Army scarcely mention Vietnam and the regular Army is as unprepared now as it was then to fight a protracted counter-insurgency campaign; the other Services are equally unprepared for low intensity conflict. Ironically, while many feel that more could be taught of Vietnam in the various military and tactical training schools, interest in the Vietnam War is sharply increasing in American civilian universities. For the Army, whose focus has always been on the Central Front, it appears that Vietnam was but a large bump on the road to Europe. Many Army officers have stated that Vietnam has remained unstudied because senior officers felt that in doctrinal terms the Asian experience was irrelevant to Europe and may indeed be counter-productive.

Any difficult war would have revealed the serious, perhaps

potentially fatal, weaknesses in the US method of 'managing' war. After a relatively short time in Vietnam some frightening problems became manifest in the Army, including the disintegration of units, localised mutiny, the increasing murders of officers and NCOs by their men, drug abuse, black marketeering and corruption, and more. These were not the marks of a motivated, well-trained and well-led army.

If one accepts the traditional premise that when an endeavour fails, the leaders shoulder the blame, then the soldiers, who never lacked for bravery, were failed badly by their civilian and military leaders. Corruption which allowed the richer to avoid conscription, the shallow training of officers and men, the rapid turnover of soldiers, the six-month tour for battalion commanders, the unprecedented interferences in small-unit tactical operations by senior officers and civilians, the infatuation with high technology and the acquiescence in disaster by the generals and admirals, all ensured that an army under stress would eventually show the cracks. Westmoreland has said that he saw the communists as 'bully boys' with crowbars, ready to wreck the building, and his job was to keep the bully boys away while the building was strengthened.[31] But it was not only the building which began to reveal its flaws — the building's protectors, the US Army, began to decay in the field after only three years of combat.[32] For the Americans, signing the Paris Peace Accord of 1973 became in some ways an imperative for their own survival.

Although not a novel assertion, the dominant feature of the war arguably was not the strategy, violation of the principle of unity of command, micromanagement, Service selfishness, or the like — it was a combination of high-level amateurishness and an accompanying massive breakdown of professional and personal integrity. All else flowed from this. This was characterised by countless incidents throughout the war; these ranged from highly publicised happenings such as the My Lai atrocities (March 1968), where virtually everyone involved, from generals on down, lied to cover it up (it was finally revealed by a sergeant), through the lying about body counts to lesser known incidents. The 7th Air Force Commander, for example, went around the world claiming that a certain bombing system was achieving significant results, yet was unaware that members of his staff asserted that they were altering the figures (by not including bad misses) to fortify his claims.

The effect of Vietnam on the Army was as destructive as it was predictable. Even among those in Vietnam who were not in the field with the Army, there was a feeling that something terrible was happening to the Army. The appearance of the troops, their comments and conversation all pointed to a systematic break-down of sorts. Years later, this was quantified to an extent in a controversial study:

> In the final years of the Vietnam conflict unambiguous signs of rapid internal military decay appear for which we can find no proper parallels anywhere. These indicators include the following: rising desertion rates over a ten year period with a great acceleration toward the end of the conflict, and by 1971 far exceeding World War II and Korean rates; mutinous outbreaks in combat units; attempted and actual murder of officers by their troops in rising numbers; and a drug addiction plague of vast proportions especially in the last four years of the war.[33]

These conditions resulted from 'a moral decay among leadership tending to elicit contempt from their men. . . '[34]

This study by Gabriel and Savage charged emotions and caused much debate in the forces. Many Army officers denied any causal relationship between the officers and soldiers, as described above. The officers criticised by Gabriel and Savage were of a higher rank and position than those who were murdered, and those incidents generally occurred in rear support units such as supply or tranportation units or in units pulled back from combat; those higher officials remained unscathed. The lower-ranking officers — usually captains, lieutenants and non-commissioned officers — paid the price for conditions not of their making.

While widespread careerism and 'ticket-punching' did much to enhance individual officers' promotions, they wounded the Army. Units became the unhappy bearers of non-traditional commanders, who were just passing through for a few months on their way to bigger and better things. But there was another reason for the Army's state, not often remembered these days. Perhaps far more than the Vietnamese communists, an American Secretary of Defense had earlier laid the groundwork for the cracking of the Army. This was Robert McNamara's infamous

'Project 100,000', a scheme by which the Services — particularly the Army — became a dumping ground for 100,000 young men of less than average intelligence, criminal and other anti-social and drug-related backgrounds. The Army was to be a sort of grand 'borstal' whose discipline would straighten out these misfits. As expected, unit discipline suffered, for during this period civil legal standards were being increasingly applied to the military forces, whose commanders over the years saw their traditional ability to punish gradually eroded. The Services spent years dealing with these poor soldiers, many of whom were expelled from the forces with less than honourable discharges.

In the United States, the concept of *noblesse oblige* towards the nation's defence is virtually completely lacking and the war was fought by the less privileged classes who were conscripted in a decidedly undemocratic manner. The richer middle and upper class youth, sustained by their families, avoided the draft (conscription) by several means — from long stays at college to disappearance to Canada or Sweden — and the war in Vietnam grew gradually into a class war fought by underprivileged teenagers who often hated the Regular Army (which, according to researchers, suffered fewer casualties in the war by apparently proportionately fighting less).[35] Battalion commanders, for instance, were exposed to combat for only six months while conscripted privates had to face combat for 12 months.

One notable characteristic of this war, not often described, was what amounted to an adversary relationship which existed to varying degrees between all layers of the US command. The young conscripted soldiers often despised the regulars; those in the field had little use for the weighty theatre headquarters staffs; the headquarters staffs resented the inordinate amount of time they had to devote to the various bureaucracies in Washington rather than on the war itself. Then, as now, it was often impossible to get the truth to the senior generals. If a staff officer was scheduled to brief a general, he first had to brief several layers of supervisors in between. Each colonel or general, regardless of his lack of detailed knowledge, altered the briefing as he saw fit, so that when the senior general finally received the message it may or may not have contained what the responsible 'action officer' really wanted to say. The generals in turn often packaged the information as they saw fit. Thus many decisions were undoubtedly based on faulty information or naked political expediency.

Above the company-grade officers, leadership in battle was often from the rear or from above, with increasing interference in small-unit actions by senior officers hovering in layers of helicopters above the battle. One infantry officer reported that during a difficult fight his problems were compounded by overlapping queries, interference and conflicting gratuitous guidance offered by a battalion commander hovering above in a helicopter, and above him, successively, the brigade and division commanders in their helicopters.[36] Micromanagement became — and remains — a curse to effective mission accomplishment.[37]

The problem lay in the upper echelons of American political and military leadership. This feeling was shared by senior Vietnamese officials who recognised the inability of the higher American officers and civilians to come to grips with Vietnam. The Vietnamese stated that the younger US officers, in contrast to their superiors, were eager to learn about Vietnam.

As Gabriel and Savage later wrote, 'in the end, factors associated with military decay focus on the officer corps, a corps unsure of itself and its standards of conduct, finally unable to enforce basic discipline, overmanaged with superfluous staff and held in contempt by their troops'.[38] Many would qualify this assertion to state that this applies generally to higher ranking officers.

One repeatedly sees the statement that US units never lost a battle.[39] But Westmoreland's memoirs cite an account of the rout of an entrenched US unit by an enemy force less than one fifth its size[40] and many small US camps, bases and units were indeed overrun and wiped out during the course of the war. The oft-written statement should be clarified to read that 'no large US unit, massively supported by helicopters, fixed-wing aircraft, artillery and sometimes naval gunfire was defeated by (generally smaller) enemy units not possessing similar fire support'. That would be a more correct statement, and if the communists had not yet won the war by 1973, neither had the US, whose forces were rarely able to hold the field after a battle.

A fundamental problem with the inability to achieve lasting results in Vietnam was the relative scarcity of fighting men in the Army. Of over half a million soldiers in Vietnam during the peak years of 1968–9, the US could field only about 80,000 actual fighters — such is the penalty an army pays for technological sophistication. This remains a problem. Given the inability of the United States to match in manpower its two main potential adversaries, the Soviet Union and China, the US Armed Forces have been forced to rely more heavily on technology and women to make up its deficiencies; both avenues of redress impose penalties in cost and in fighting capability, the one requiring ever more service and support troops while the other group is prohibited by law from direct combat. Furthermore, a tenth of the women soldiers are pregnant at any given time.

Today the United States and its Army is faced with a growing and potentially fatal challenge in its own hemisphere. In Central America several communist insurgencies are flaring simultaneously with varying degrees of success. While El Salvador is most publicised, it is but a piece of the whole, other insurgencies occurring in Honduras, Guatemala and Nicaragua (the latter against the Marxists in power). For the leftists, the grand prize is Mexico; should the Western hemisphere be split by a 'red belt' around its middle, Mexico, with its internal social and economic structures, is a case of an insurgency waiting to happen. Other serious Marxist insurgencies are occurring in Colombia and Peru.

In El Salvador, 6000–7000 Marxist revolutionaries, employing the classic phased approach to revolutionary war, have entrenched themselves in the remote mountainous areas as they attempt to build their infrastructure and conduct guerrilla warfare. El Salvador has little money, and even less experience, to provide for pacification efforts, though attempts are being made in selected areas. American advisers are forbidden by their own government from accompanying the Salvadoran Army on operations. Ironically, while the US has emphasised mobile, small-unit tactics in El Salvador, in selected regions the Salvadoran government appears to have rejected US advice and is continuing to mount the big-unit operations so generally unsuccessful in Vietnam.

If one accepts that the government forces need a numerical superiority of ten to one over the insurgents, then the government of El Salvador is doomed to failure, as the ratio is nearer three or four to one. The small number of American advisers, a total of 55, are all restricted to the city limits of the capital, San Salvador. Attempts to increase this number are met with reoccurrences of the 'Vietnam Syndrome', as congressmen, senators, newsmen and others raise the spectre of another

Vietnam. Another small group of advisers is beginning to train Salvadoran and local soldiers in adjacent Honduras. But given the steadfastness with which the Soviet bloc supports its allies and clients, combined with the volatility of the Americans and a seeming inability of the targeted regimes to reform, the likelihood remains that a series of Marxist states will eventually be established in Central America.

But these smouldering wars on America's doorstep have served in part to refocus the Army's attention on low intensity conflict. For nearly a decade the lessons of Vietnam remained unstudied by the Army as its planners returned to assessing the relative forces and strategies of the European battlefield, the Army's major area of interest. In the early 1980s the Airland Battle doctrine was promulgated, a doctrine concerned with the Central Front but applicable worldwide. This doctrine postulates attacking the enemy forces in depth, synchronising all combat arms at hand (to include those of other Services). Many Army officers state that the development of this doctrine was delayed by a decade because of the US involvement in Vietnam.

Insurgency and counter-insurgency are not mentioned in the US Army's fundamental manual — Field Manual 100–5, 'Operations' — described as 'the Army's keystone How to Fight Manual'. Periodic revisions of this basic manual have placed less and less emphasis on counter-insurgency, and in the latest revision (20 August 1982) any references to counter-insurgency have been omitted. The study of counter-insurgency was left largely to the Special Forces, whose main role and *raison d'être*, strictly speaking, is to *be* the guerrilla and to organise insurgencies, not to fight them.

But the US Army has acknowledged the fact that limited and unconventional wars are the wars of today — what Roger Trinquier calls 'Modern Warfare' — and that the massive conventional frontal war against the communist armies in Europe is the least likely of wars.[41] But because of the great impact on US national security of the European theatre, the Army had until recently chosen to neglect the unconventional wars in order to concentrate its resources on the more conventional scenario in Europe. After Vietnam, the light infantry brigades and divisions gained weight, became mechanised and were rendered more roadbound.[42]

There are, however, several important Army initiatives under way which grew out of the Army's performance in Southeast Asia. Under the direction of the former Chief of Staff, General E.C. Meyer, the Army began to look at reorganising into a regimental system as found in the British Army (and in the US Army in an earlier age). Project 'Cohort' was established to study this and other steps to reduce the present levels of organisational turbulence caused by frequent personnel transfers and unit reorganisations. Brigade and battalion commanders, who during the Vietnam War rotated in and out of troop command every six months on a revolving door basis, were to remain in command for 30 to 36 months. (This was reduced to 24 months under the new Chief of Staff, General Wickham, in 1984.) Units were trimmed to fit resource realities; for example, the size of a squad, which had ranged from nine to eleven men, was fixed at nine men, and howitzer crew size at nine (from ten).

In the last two years (1983–4), the Army has worked out a low intensity warfare doctrine, and drafts of this paper are being studied. This will affect the Army's special-purpose forces geared to this end of warfare's spectrum: the Special Forces, Rangers, Psychological Operations units and Civil Affairs personnel. There was some initial disagreement on the low intensity doctrine, as some regular Army officers believed that the Special Forces were still influenced by the experience in Vietnam and by older 'Vietnam hands' who fought their own war in the Highlands away from the population. This is a matter of continuing debate.

The major Army reorganisation now occurring reflects another turn of the wheel with the recreation of light infantry divisions, restructuring of the heavy divisions and the contemplated conversion of an infantry division to a 'high technology' division. With completion of this force restructuring, the Army hopes to be better prepared to fight any war from the high to the low ends of the spectrum of intensity.

Finally, the end of conscription — which came about from necessity born of domestic political unrest — may preclude the kind of resistance to the war found within the Army itself, and increase the professional competence of the soldiers. The abandonment of conscription at the end of the war was not a fundamental course correction; conscripts have fought well in similar situations around the world. But the end of the draft initially caused its own problems as less well-educated soldiers were initially recruited, although this trend has been reversed in recent years.

The key question remains — was the Vietnam War 'winnable'? All things considered, the answer must be 'probably not'; the communists, unlike the Americans, would have waited a century for this victory, while the fragmented South Vietnamese were under siege equally by the communists and the Americans, the latter incessantly trying to clone the Vietnamese political system into an American image. Few South Vietnamese or Americans appear to have understood the social and political bases for the war. And as long as the United States operated under self-imposed restraint in denying its regular ground forces access to the enemy's vital resupply and reinforcement routes and rest areas in Laos, any hopes of fighting a conventional war were unlikely to be realised, even if that had been an appropriate strategy.

The final result in Vietnam produced predictable conclusions in the United States: the politicians blamed military ineptitude for the disaster in Vietnam, while the generals blamed everyone but themselves in declaring the war a draw; the younger officers saw the war as a humiliating defeat brought on by the ineptitude of senior officers[43] and politicians (many stated that they could not distinguish between the two); the South Vietnamese saw an American betrayal; the rest of the world noted the humiliation and apparent weakness of the United States. Lacking strong political control, an achievable overall aim, an appreciation of the nature of the threat and a workable response to it, the world's strongest power had been defeated by 'puny men of a puny nation'.[44] It was a failure of counter-insurgency which deserves close examination by other nations involved in similar campaigns.

Notes

1. John E. Beebe Jr., 'Beating the Guerrilla', *Military Review*, vol. XXXV, December 1956, p. 6.

2. BDM Corporation, *A Study of Strategic Lessons Learned in Vietnam*, vol. VI *Conduct of the War*, Book 2 'Functional Analyses' (McLean, Virginia, 1980), p. II-1.

3. William Westmoreland, *A Soldier Reports* (Doubleday, New York, 1976), p. 161.

4. Maxwell D. Taylor, *Swords and Plowshares* (Norton, New York, 1972), p. 316. See also W. Scott Thompson and Donaldson Frizzel (ed.), *The Lessons of Vietnam* (Crane, Russak, New York, 1977), p. 186.

108   *The American Army*

5. Taylor, *Swords and Plowshares*, p. 311.
6. United States National Archives, Record Group 273, Records of the National Security Council, National Security Study Memorandum (NSSM) 1, 29 January 1969.
7. John J. Tolson, Vietnam Studies, *Airmobility, 1961–1971* (Department of the Army, US Government Printing Office, Washington DC, 1973), p. 124.
8. Bruce W. Watson and Peter M. Dunn (ed.), *Military Lessons of the Falkland Islands War: Views from the United States* (Westview Press, Boulder, Colorado, 1984), p. 75.
9. Thompson and Frizzel, *Lessons of Vietnam*, p. 243.
10. Ibid.
11. NSSM 1, CIA Response, 7 February 1969.
12. Personal observation by the author, Saigon, 1972. The British were to have this problem in the Falklands campaign of 1982, see Watson and Dunn, *Lessons of the Falkland Islands War*, p. 72.
13. Joseph A. McChristian, Vietnam Studies, *The Role of Military Intelligence, 1965–1967* (Department of the Army, GPO, Washington DC, 1974), p. 22.
14. Ibid., p. 164.
15. Ibid.
16. BDM Corporation, *Strategic Lessons Learned in Vietnam*, vol. VI, Book 2, pp. 9-29; also Thompson and Frizzel, *Lessons of Vietnam*, p. 143.
17. McChristian, *Military Intelligence*, p. 164.
18. Bernard W. Rogers, Vietnam Studies, *Cedar Falls — Junction City: A Turning Point* (Department of the Army, GPO, Washington DC, 1974), p. 78.
19. Thompson and Frizzel, *Lessons of Vietnam*, p. 216.
20. Ibid., pp. 216, 247.
21. NSSM 1, 21 January 1969.
22. Guenter Lewy, *America in Vietnam* (Oxford University Press, New York, 1978), pp. 280-1.
23. Frank Snepp, *Decent Interval* (Random House, New York, 1977), p. 567.
24. Robert Komer, *Bureacracy Does its Thing: Institutional Constraints on US-GVN Performance in Vietnam* (Rand Corporation, Santa Monica, California, 1972).
25. Remarks by US Army historians, Vietnam Historians Workshop, US Marine Corps History and Museums Division, Washington DC, 9 May 1983.
26. Thompson and Frizzel, *Lessons of Vietnam*, p. 250.
27. Ibid., pp. 246-7.
28. Ibid.
29. Ibid., p. 253.
30. Jack Shulimson and Charles Johnson, *US Marines in Vietnam, The Landings and the Buildup, 1965* (USMC, GPO, Washington DC, 1978), p. 135.
31. Westmoreland, *A Soldier Reports*, p. 145.
32. Major Robert A. Doughty, *The Evolution of US Army Tactical Doctrine, 1946–1976* (Combat Studies Institute, Fort Leavenworth, Kansas, August 1979), p. 40.
33. Richard Gabriel and Paul Savage, *Beyond Vietnam: Cohesion and Disintegration in the American Army* (International Studies Association, 16th Annual Convention (n.d.), Washington DC), p. 16.
34. Ibid., p. 23
35. Ibid., p. 37
36. See photographs in Rogers, *Cedar Falls — Junction City*: p. 22 shows an assistant division commander, a one-star general, taking to his helicopter to 'direct his troops in battle'; p. 28 depicts another general checking 'the location of a patrol'.
37. See HEW Report on DDPMA (Deputy Director Politico-Military Affairs)

OJCS (Organisation of the Joint Chiefs of Staff), Washington DC, 14 February–4 March 1980. This was a survey of officers of all Services in grades of major through colonel serving in the OJCS Plans and Policy Directorate (J-5). The survey revealed a remarkably deep contempt for the generals and admirals and their poor leadership and managerial abilities; scores of officers were interviewed. Most of the generals and admirals referred to were promoted to higher ranks.

38. Richard Gabriel and Paul Savage, *Crisis in Command: Mismanagement in the Army* (Hill and Wang, New York, 1978), p. 96.

39. Colonel Harry G. Summers, *On Strategy: The Vietnam War in Context* (US Army War College, Carlisle Barracks, PA, 1981), p. 1. See also Doughty, *Evolution of US Army Tactical Doctrine* pp. 39–40.

40. Westmoreland, *A Soldier Reports*, p. 368.

41. Doughty, *Evolution of US Army Tactical Doctrine*, p. 40.

42. BDM Corporation, *Strategic Lessons Learned in Vietnam*, Omnibus Executive Summary, 1980, p. VIII-2.

43. See Jeffrey Record, 'Why Our High-Priced Military Can't Win Battles', *The Washington Post*, 29 January 1984, p. D1. To varying degrees there persists a belief, in and out of the services, that the Military Departments have become federal bureaucracies.

44. Walter Walker, *East of Katmandu* (Leo Cooper, London, 1976), p. xvii.

References

Baratto, D. 'Special Forces in the 1980s: A Strategic Reorientation', *Military Review*, vol. LXIII, March 1983

BDM Corporation, *A Study of Strategic Lessons Learned in Vietnam* (multi-volume series, McLean, Virginia, 1979–80)

Beebe, J. 'Beating the Guerrilla', *Military Review*, vol. XXXV, December 1956

Blaufarb, D. *The Counter-Insurgency Era: US Doctrine and Performance, 1950 to the Present* (Free Press, New York, 1977)

Charlton, M. 'Many Reasons Why', *The Listener* (British Broadcasting Corporation, London), 22 September 1977

Cooper, C. *et al.The American Experience With Pacification* (3 vols. Institute for Defense Analyses, Arlington, Virginia, 1972)

Doughty, R. *The Evolution of US Army Tactical Doctrine, 1946–1976* (Combat Studies Institute, Fort Leavenworth, Kansas, 1979)

Duncanson, D. *Indochina: The Conflict Analysed* (Conflict Studies, London, 1973)

Eckhardt, G. *Command and Control, 1950–1969*, Vietnam Studies series (hereafter VS). (Department of the Army, Government Printing Office (GPO), Washington DC, 1974)

Gabriel, R. and Savage, P. *Beyond Vietnam: Cohesion and Disintregation in the American Army* (International Studies Association, 16th Annual Convention (n.d.), Washington DC)

Gabriel, R. and Savage, P. *Crisis in Command: Mismanagement in the Army* (Hill and Wang, New York, 1978)

Haslorff, J. 'Vietnam in Retrospect: An Interview with Ambassador Frederick E. Nutting, Jr', *Air University Review*, vol. XXV, no. 2, Jan-Feb. 1974

Johnson, L. *The Vantage Point: Perspective of the Presidency, 1963–1969* (Popular Library, New York, 1971)

Karnow, S. *Vietnam* (Viking Press, New York, 1983)

Kelly, F. *US Army Special Forces, 1961–1971* (VS) (Department of the Army,

GPO, Washington DC, 1973)

Kinnard, D. *The War Managers* (University Press of New England, Hanover, New Hampshire, 1977)

Kissinger, H. *White House Years* (Little, Brown, Boston, MA, 1979)

Komer, R. *Bureaucracy Does Its Thing: Institutional Constraints on US-GVN Performance in Vietnam* (Rand Corporation, Santa Monica, California, 1972)

Larson, S. and Collins, J. *Allied Participation in Vietnam* (VS) (Department of the Army, GPO, Washington DC, 1975)

Lewy, G. *America in Vietnam* (Oxford University Press, New York, 1978)

McChristian, J. *The Role of Military Intelligence, 1965–1967* (VS). (Department of the Army, GPO, Washington DC, 1974)

Momyer, W. *Air Power in Three Wars* (GPO, Washington DC, 1978)

Ney, V. *Evolution of the United States Field Manual. Valley Forge to Vietnam.* Combat Operations Group Memorandum (CORG-M-244), January 1966

Nixon, R. *The Memoirs of Richard Nixon* (Grosset and Dunlap, New York, 1978)

Parker, W. *US Marine Corps Civil Affairs in I Corps, Republic of Vietnam April 1966 to April 1967* (GPO, Washington DC, 1970)

Race, J. *War Comes to Long An: Revolutionary Conflict in a Vietnamese Province* (University of California Press, Berkeley, CA, 1972)

Rogers, B. *Cedar Falls — Junction City: A Turning Point* (VS) (Department of the Army, GPO, Washington DC, 1974)

Scoville, T. *Reorganizing for Pacification Support* (GPO, Washington DC, 1982)

Sharp, U.S.G. and Westmoreland, W. *Report on the War in Vietnam* (GPO, Washington DC, 1968)

Sharp, U.S.G. *Strategy for Defeat* (Presidio Press, San Rafael, CA, 1978)

Shulimson, J. and Johnson, C. *US Marines in Vietnam, The Landings and the Buildup, 1965* (USMC, GPO, Washington DC, 1978)

Shulimson, J. *US Marines in Vietnam — An Expanding War, 1966* (GPO, Washington DC, 1982)

Snepp, F. *Decent Interval* (Random House, New York, 1977)

Stolfi, R. *US Marine Corps Civic Action Efforts in Vietnam, March 1965–March 1966* (GPO, Washington DC, 1968)

Summers, H. *On Strategy: The Vietnam War in Context* (US Army War College, Carlisle Barracks, PA, 1981)

Taylor, M.D. *Swords and Plowshares* (Norton, New York, 1972)

The Vietnam Experience (multi-volume series on the Vietnam War, The Boston Publishing Company, Boston, MA, 1981–3)

Thompson, R. *No Exit From Vietnam* (David McKay, New York, 1969)

Thompson, R. *Peace Is Not At Hand* (Chatto and Windus, London, 1974)

Thompson, W.S. and Frizzel, D. (ed.), *The Lessons of Vietnam* (Crane, Russak, New York, 1977)

Tolson, J.J. *Airmobility, 1961–1971* (VS) (Department of the Army, GPO, Washington DC, 1973)

US Army. *DA Pamphlet No. 550–100, Handbook of Counterinsurgency Guidelines for Area Commanders* (1966)

US Army. *Field Manual 100–5: Field Service Regulations: Operations* (Washington DC, 1962, 1968, 1976, 1982)

United States National Archives, Record Group 319, Records of the Army Staff (G-3); Indochina (The 'Gavin Plan'), 1954

United States National Archives, Record Group 273, Records of the National Security Council, National Security Study Memorandum 1, 1969

United States, Department of Defense. *The Pentagon Papers* (Senator Gravel Edn, 4 vols. Beacon Press, Boston, MA,1972)

Walker, W. *East of Katmandu* (Leo Cooper, London, 1976)

Walt, L. *Strange War, Strange Strategy* (Funk and Wagnalls, New York, 1970)
Westmoreland, W. *A Soldier Reports* (Doubleday, New York, 1976)

4 THE LATIN AMERICAN EXPERIENCE: THE TUPAMAROS CAMPAIGN IN URUGUAY, 1963–1973

F.A. Godfrey

The Latin American states are as different, one from another, as are the states of Europe. They vary enormously in size and in population; their dominant cultural influences may be Spanish, Portuguese, British, Dutch or French; the racial composition of their people is frequently diverse; geographic and climatic conditions also stand in stark contrast — yet the European student of the region often tends to see them as set in the same mould.

This tendency is probably a result of the fact that the states of Latin America, almost without exception, suffer an unenviable reputation for political instability, based upon a seemingly endemic restlessness since their independence from colonial rule, mostly in the nineteenth century. Many explanations have been offered as to why this should be the case, but few provide complete or even satisfactory answers. Some commentators blame the disputed frontiers of a colonial heritage; although this has undoubtedly produced tension between states, it has rarely led to open conflict or instability, the military clashes between Peru and Ecuador (1976–7 and 1981) and the Falklands War between Britain and Argentina (1982) being the exceptions. Nor can one seriously attribute a high incidence of violence, as the casual observer might, to an excitable national or ethnic character. More realistically, persistent economic problems — apparent despite the existence of vast natural resources — an unequal distribution of income and a record of continuous interference by outside bodies, both economic and political, have played a major part in the frequent changes of government, although once again, these problems need not necessarily lead to violence. As in so many other aspects of the human condition, the instability and conflict so noticeable throughout the region has its roots in a unique mix of diverse considerations.

But whatever the reasons, it is worth noting that violence in

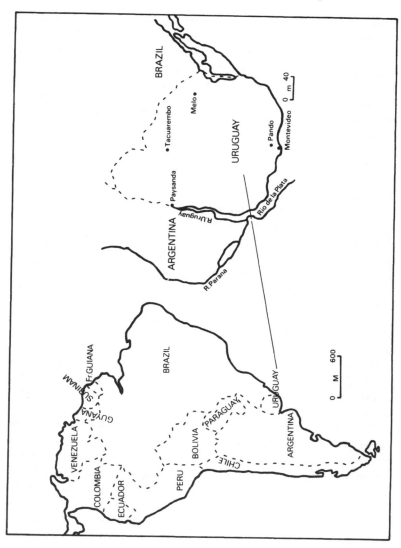

Map 4.1:
Latin America

Latin America has rarely manifested itself in conventional military operations between sovereign states since 1945, even though this tended to be a characteristic of earlier periods. Argentina, for example, was involved in no war between 1869 and the disastrous Falklands enterprise of 1982, her entry into the Second World War in 1945 being entirely on a diplomatic level, and only two Latin American states — Brazil and Mexico — were actively involved in the war against the Axis powers. With the exception of the relatively minor frontier clashes already mentioned, the only indigenous conflict approaching major hostilities in recent years was the bloody but short-lived 'Football War' between Honduras and El Salvador in 1969. Where violence has erupted, it has invariably taken the form either of usurping existing government authority by means of a *coup d'état*, through which the armed forces have either seized power for themselves or come to exert a major influence over the politics of the state, or in the development of insurgency aimed at securing particular social or political demands. The incidence of coups has been high — some 81 coups or attempted coups were recorded in Latin America between 1945 and 1971 alone[1] — but there have also been a wide variety of insurgency campaigns to provide the Security Forces with alternative employment.

Historically, insurgent movements in Latin America have been fostered by diverse groupings of disaffected people. Young officers of the armed forces have often been involved and, indeed, one of the leading Brazilian urban guerrillas of the late 1960s, Carlos Lamarca, was a former army officer. Elements of the educated middle classes have also played their part while, in more recent years, students have taken a major role, as have members of militant trades unions. Among what might be termed the more 'traditional' movements seeking to bring armed force to bear against government, aims were of a more limited nature for the most part. Attempts were often made radically to change the thrust of a government's policies or even to seize power in order to implement a dramatic change of course, but on the whole these 'traditional' movements did not seek to foment a revolution which would completely destroy the existing structures of the state and replace them with a new order.

In the early 1960s, all this was to change. The success of Fidel Castro on Cuba, where he achieved power in 1959, had enormous repercussions for governments throughout Latin

America. The insurgent movements against which they now had
to wage war were of a completely different kind. New groupings
motivated by different ideologies were soon to prove themselves
a far greater menace than anything hitherto experienced. What
was more, these groups were to receive support, not always
physical but frequently vocal, from external sources, and this
gravely exacerbated the difficulties of governments in their
attempts to deal with the menace. Thus, in the wake of Castro's
victory over Fulgencio Batista, there were similar risings in
Colombia (1961), Guatemala and Ecuador (1962) and Peru
(1963), a process which was to culminate in Ernesto 'Che'
Guevara's ill-fated expedition to Bolivia in 1967.

As was the case in Castro's campaign in Cuba, the environment of these new insurgency organisations was primarily the
countryside, although it should be noted that Castro's landing
from the *Granma* in December 1956 was originally supposed to
coincide with urban risings in the Oriente province and that it
was the collapse of the latter that forced the *Fidelistas* up into the
Sierra Maestros mountains. Subsequently, urban groups such as
the '26 July Movement' made a significant contribution to
Castro's ultimate success by tying down an estimated 15,000
troops — approximately half of Batista's army — even if
individual actions such as the rising at the Cienfuegos naval base
(September 1957) or the general strike in Havana (April 1958)
were failures. The urban contribution was, however, conveniently forgotten by Guevara and Régis Debray, the French
Marxist philosopher, in their formulation of the 'foco' theory of
rural insurgency, although both acknowledged that urban operations could assist the success of rural action. In being related
directly to the experience of Cuba, 'foco' also rejected the
emphasis in prevailing rural guerrilla warfare theory — notably
that of Mao Tse-tung — on the necessity for a prolonged period
of political preparation prior to the commencement of insurgency. For Guevara and Debray a minimum level of popular
discontent with a government would provide a sufficient political
base for the military actions of a small elite band of fast-moving
and hard-hitting guerrillas. The failure of a government to
suppress the guerrillas would force it into repression affecting the
entire population and thus exposing the corrupt nature of the
system as a whole to the latter. The guerrillas would then prove
the 'foco' or focus for wider insurrection.

Historically, of course, there was nothing new in an emphasis on rural operations. Most guerrilla organisations have sought to operate against government forces from the relative safety of rural strongholds, where the military weakness of the movement compared to the strength of government forces is less important. Small groups striking against government forces spread too thinly on the ground can exact a toll out of all proportion to the ratio of forces in the confrontation. As Robert Taber, the American journalist who fought with Castro, observed, Batista's deployment of over 5000 troops to hunt down the small band of Castro's guerrillas defied the 'simple arithmetic' of these men covering an area of over 3885 sq.km (1500 sq. miles) of heavily forested terrain.[2] In Latin America, however, the rural 'foco' campaigns did not succeed, and later guerrilla organisations drifted towards urban action (although there was one early example of urban-based insurgency in Venezuela between 1962 and 1965).

There were a number of reasons for this fundamental change in revolutionary strategy besides failure in the countryside. Urban terrorism had appeared to work in both the EOKA campaign against the British authorities in Cyprus (1955–9) and the *Front de Libération Nationale* (FLN) campaign against the French in Algeria (1954–62), although the ultimate outcome of the 'Battle of Algiers' in 1957 and of the campaign in Venezuela in 1965 should have served due warning that urban insurgency could prove equally disastrous to the guerrilla. Rather more significantly, it was a recognition of the fact that the great majority of the population of Latin America no longer lived in the countryside. By 1967 at least 50 per cent of the population of every South American state, except Peru, inhabited urban areas, with some, such as Uruguay and Argentina, registering well over 70 per cent of the population as urbanised. With high unemployment, high inflation and a concentration of such a large proportion of a relatively youthful population in urban slums and 'shanty towns', the widespread sense of deprivation was ripe for exploitation amidst a seemingly general radical upsurge in the Western world.

It should be noted, however, that the 'theory' of urban guerrilla warfare as it emerged in the late 1960s and 1970s did not necessarily rule out rural action. Admittedly, as in Guatemala in 1967, the movement of the guerrillas to the urban environment was an expression of failure in the countryside, but the best known theorist of urban insurgency, Carlos Marighela, inter-preted such urban action as facilitating rural insurgency by drawing the armed forces into the cities. By contrast to Marighela in Brazil, the Tupamaros of Uruguay attempted to reopen a rural front in 1971–2 to take pressure off their operations in the cities. Similarly, *Ejército Revolucionario del Pueblo* (ERP) opened a rural campaign in Argentina's Tucaman province in 1975 which necessitated the deployment of the Army's 5th Brigade.

For the most part, however, urban warfare theorists such as Marighela and the Spaniard, Abraham Guillen (who influenced the Tupamaros), believed that society and government could best be paralysed by guerrilla action in the main centres of population and commerce. Not unlike the 'foco' theory, the urban guerrillas would be a small elite band, organised in a cellular structure for better security, whose 'armed propaganda' would provoke a government reaction out of all proportion to the numbers of the insurgents. In the words of Marighela, guerrilla successes would 'force those in power to translate the political situation of a country into a military situation'. This, in turn, would 'alienate the masses who, from then on, will revolt against the army and the police and thus blame them for this state of things'.[3] For the urban guerrilla, the media would be the vital weapon; 'armed propaganda' such as bank raids, street ambushes, spectacular kidnappings and sabotage being designed to win maximum publicity. Moreover, the victims of guerrilla action would not just be the 'establishment', but also foreign multinationals, with the subsidiary aim of weakening the economy by driving foreign capital out of the country.

A number of the guerrilla groups that emerged in the 1960s and 1970s — but especially the urban groups of the latter decade — had informal links with each other as well as with external powers such as Cuba. In February 1974, for example, ERP from Argentina, Bolivia's *Ejército de Liberación Nacional* (ELN), Chile's Miristas and Uruguay's Tupamaros issued a joint communiqué, and together with the Frepalina of Paraguay, these groups formed a *Junta de Co-ordinacion Revolucionaria* with headquarters in Paris in the same year. But, if the guerrillas enjoyed a degree of external support, it must also be recognised that Latin American Security Forces benefited enormously in their counter-insurgency campaigns from the assistance of the

United States. Indeed, the contribution of the US to the defeat of the rural insurgencies of the early 1960s was particularly significant.

The US has, of course, a long record of involvement in Central and South America, from the Monroe Doctrine of 1823 onwards. Between 1830 and 1945, US forces were sent into Latin American states on no less than 70 occasions[4] and there has been little hesitation in intervening whenever US interests have been perceived as being under threat. Since 1945, such intervention has been direct — as in the case of the Dominican Republic (1965) and Grenada (1983) — and indirect — as in the US involvement in the coups in Guatemala (1954) and Chile (1973). But the support of indigenous armed forces is clearly both preferable and cheaper than the deployment of US Marines, President John F. Kennedy in particular being associated with the development of a global American doctrine of counter-insurgency. Conceived as a response to Nikita Khrushchev's declaration of support for 'wars of national liberation' in January 1961, the doctrine was designed to prevent the advance of communism by means other than direct US military intervention. Thus a programme of assistance to win over the allegiance of indigenous populations would be combined with military advice and technological and logistical support for Security Forces in threatened states. All the US armed forces developed special counter-insurgency capabilities, while Kennedy significantly advanced the status of US Special Forces (the 'Green Berets'), when he visited them at Fort Bragg in October 1961. A large new infrastructure of military and civilian agencies was evolved to bring reality to the 'Alliance for Progress'.

Within the context of Latin America, states thought to be at risk, such as Colombia, Peru, Venezuela and Guatemala, received US missions to assist in developing internal defence plans. Green Berets of the 8th Special Forces Group based on Fort Gulick in the Panama Canal Zone, became directly involved in operations against guerrillas. Some 52 anti-subversive missions were carried out in 1965 and possibly over 400 between 1962 and 1968. It has been claimed, for example, that over 1000 Green Berets were operating in Guatemala from 1966 and certainly 28 US servicemen were killed there between 1966 and 1968. The US also undertook bombing missions over Guatemala in 1967 while, in the same year, training camps were established in both

Nicaragua and Bolivia.[5]

Quite often, such assistance was required because Latin American armed forces were not only comparatively small but totally untrained in counter-insurgency. A case in point is the assistance given to the Bolivian Army. Once the presence of Guevara's guerrillas had been revealed in March 1967, the US hurriedly established a training camp under Major 'Pappy' Shelton and within four months the Bolivians were able to deploy the first elements of the newly-raised 2nd Ranger Battalion. The full battalion of 650 men had been deployed by September, sealing Guevara's fate: it has often been alleged that two members of the US Special Forces were present when Guevara was captured and killed. Training was also given at Fort Gulick. In 1966, for example, some 1323 Latin American military personnel undertook courses in the US and 4692 attended courses given by the US forces elsewhere than mainland USA. It has been calculated that over 20,000 Latin American military personnel passed through courses in the Panama Canal Zone between 1962 and 1970.

Of course, counter-insurgency is not just a matter for the armed forces and it has usually been the police that have borne the initial brunt of most insurgency campaigns worldwide. In Argentina the Army was not directly involved in operations against urban guerrillas until the autumn of 1975, although its involvement increased significantly after the military coup of March 1976. Similarly, the Army assumed responsibility for the campaign against the Tupamaros in Uruguay only in April 1972, even though the guerrillas had been active since 1963. Thus the US also undertook the training of Latin American police forces, first at the International Police Academy in the Canal Zone and later in Washington, a six-week course being taught in the medium of the Spanish language. The principal means of assisting police training was through the US Agency for International Development (AID) which, for example, trained 3500 Latin American police officers between 1966 and 1970. The American Dan Mitrione, executed by the Tupamaros in August 1970, was an AID expert attached to the Uruguayan prefecture of police. In 1972 the Agency's Office of Public Safety was assisting 15 Latin American police forces at an annual cost of three million dollars.[6]

With the collapse of rural insurgency, the programme launched by the Kennedy Administration was effectively terminated, but the revival of guerrilla insurgency in Central America has seen increasing US involvement in recent years. This has to a large extent reproduced the earlier pattern. In the case of Nicaragua, where the Sandinista guerrillas toppled the Somoza regime in July 1979, the Central Intelligence Agency (CIA) has been implicated in the organisation of anti-government forces — the Contras. These comprise the Nicaraguan Democratic Front (FDN) based on Honduras and the Revolutionary Democratic Alliance (*Arde*) based on Costa Rica. In addition, the US was active in reviving the 1964 CONDECA alliance of El Salvador, Honduras, Panama and Guatemala in November 1983 and has undertaken large-scale manoeuvres in Honduras since mid-1983. Most recently, the CIA has been implicated in the mining of Nicaraguan ports. In the case of El Salvador, where the *Farabundo Marti* National Liberation Front has been waging guerrilla war against the government since 1977, the US has become closely involved in supplying and supporting government forces. In particular the first two of a number of planned *cadoza* or 'hunter' battalions were deployed in the summer of 1982 following training at Fort Bragg. In the spring of 1983 the US also encouraged the El Salvadoran government to announce its 'National Campaign Plan', beginning a 'hearts and minds' policy in the provinces of San Vicente and Usulutan which included the construction of new schools and clinics and the promotion of a weaving project. There has also clearly been direct involvement, CIA agents having apparently conducted missions inside Nicaragua and the first US military adviser being killed in El Salvador in May 1983.

The promotion of 'hearts and minds' in El Salvador by the US is significant since it was also one of the prime features of the counter-insurgency doctrine of the 1960s, whereby the Americans attempted to instil a sense of the need for what was then termed 'military-civic action'. In 1962, for example, the assistance given to Colombia, Guatemala and Ecuador was primarily seen in these terms, some 1.5 million dollars being allocated to Ecuador alone for improvements to roads, water supplies, health care and sanitation. Yet, in reality, many of the armed forces of Latin America already had a tradition of assisting civil development projects which stood them in good stead when insurgency began. In Colombia, the Army's officers needed little additional encouragement from the US in the field of military-civic action when insurgency commenced in 1962, introducing roads, clinics and a courteous treatment of civilians in its 'Plan Lazo' to win hearts and minds while eradicating the guerrillas. Similarly, it has been calculated that a fifth of the man-hours of the entire Bolivian Army was spent on civic action in 1963, conscripts frequently helping developments in their own villages. Armies, indeed, were generally more popular than police forces in many Latin American states. When the Army took over responsibility for the campaign against the Tupamaros in Uruguay, information denied to the police flowed into the Army's intelligence agencies. In Peru in 1965, the Army was far closer to the Indians of the interior than the intellectuals of the guerrilla 'foco' could ever hope to be. The Indians co-operated freely with the Army, although the Peruvians also utilised propaganda leaflets and airborne loudspeakers to carry their message of reward for co-operation to the interior. The campaign against Guevara in Bolivia is almost a text-book example of the success which attends the winning of hearts and minds prior to the development of active insurgency. Guevara established his group in a remote area of southern central Bolivia, relying on the local peasant population for support. The area was only thinly inhabited, but the local Indians were suspicious of even Bolivian intruders, let alone others, and they helped to track down Guevara's ill-assorted band of Cubans, Argentinians, Peruvians and Bolivians who had not even mastered the local Indian languages. The Bolivian Army was thus much more of a 'people's army' than the guerrillas.[7]

It is not often appreciated that even Batista's army on Cuba, best known for its brutality and corrupt inefficiency, at least attempted a degree of 'hearts and minds'. Colonel Barrera Perez, for example, planned to build schools and houses and to expand medical services in 1957, even opening a free kitchen serving some 300 people a day in the course of his 'rehabilitation' of the peasantry. Similarly, even the Guatemalan Army paid at least lip service to civic action, Colonel Carlos Araña Osorio attempting to improve the Army's image in Zacapa province in 1967 with his *Accion Civica Militar* office under the auspices of the 'Plan Piloto' for the socio-economic development of the northeast. Prior to the outbreak of insurgency, the Army had only organised literacy programmes, but now attempted to build schools, hospitals, roads and wells and even to provide school meals as well as better medical care and a healthier national diet.[8] Significantly, however, Araña was also associated with the emergence of right-wing death-squads such as *Mano Blanca* (MANO) or 'White Hand', *Nueva Organizacion Anti-communista* (NOA) and *Consejo Anti-communista de Guatemala* (CADEG), and it must be said that this is the more widespread popular image of the response of Latin American armed forces to insurgency. Many of the MANO and NOA gunmen appear to have been off-duty police or servicemen, while CADEG was more associated with landowners in the rural areas where Araña also established irregular militias composed of small landowners for local defence.

In present-day Guatemala, where insurgency still simmers, the responsibility for local defence is similarly devolved onto conscripted 'civilian defence patrols'. It is possible that as many as 10,000 people were killed by the Guatemalan Security Forces and death-squads in 1967–8 and certainly the involvement of the armed forces in counter-insurgency in Latin America has generally resulted in maximum use of force and institutionalised terror. In Argentina, off-duty police were implicated in death squads such as the *Alianza Anticommunista Argentine* (AAA), although its victims were frequently intellectuals or popular entertainers and rarely ERP or Montonero guerrillas. The various urban guerrilla groups in Argentina appear to have expected that direct military involvement in the campaign against them after autumn 1975 would result in more street battles and extensive cordon and search operations but not in more repression. In reality, the Security Forces made only a pretence at legality, with such units as GT33 of the Navy Engineering School (ESMA) alone being responsible for an estimated 3000 deaths.[9] It may be recalled that Lieutenant-Commander Alfredo Astiz, taken prisoner after the recapture of South Georgia in the South Atlantic by British forces in late-April 1982, was questioned in London about his involvement in the murders of two French nuns and a Swedish schoolgirl. Known as the 'Butcher of Cordoba', Astiz began his career in counter-terror at the Navy School of Mechanics. Some 18,000 suspects were detained in Argentina by the end of 1977 and it is estimated that between 10,000 and 20,000 deaths occurred in the *guerra sucia* or 'dirty war' after the Army's take-over in March 1976. Similarly, the Brazilian Army resorted to mass arrests and ruthless interrogation once it had assumed responsibility for counter-insurgency, its actions being co-ordinated by a new department in the War Ministry. Mass arrests after the kidnapping of the US ambassador, Burke Elbrick, for example, led directly to the elimination of Marighela in November 1969. There were also death-squads in Brazil, such as the *Escudiao da Morte* and *Operacao Badeirantes*, which may have been responsible for perhaps 1000 deaths between 1964 and 1970.

Of course, it is a principal aim of both the 'foco' and the urban guerrilla to provoke repression by the armed forces in order to discredit the system as a whole in the eyes of the people. Invariably, guerrillas in Latin America have, however, over-stepped the indistinct line between public sympathy and public opposition, as they did in Venezuela in September 1963 when *Fuerzas Armadas de Liberación Nacional* (FALN) terrorists attacked an excursion train, and this enabled President Romulo Betancourt to introduce emergency legislation. Similarly, an increase in Tupamaros violence in Uruguay in April 1972 enabled President Juan Maria Bordaberry to pass legislation which had been previously rejected by the General Assembly. The guerrillas made fundamental mistakes in involving the armed forces directly in counter-insurgency, as occurred in Argentina and Uruguay, where initially they had carefully refrained from attacking military personnel. In such states, the repression resulting from direct military involvement proved so severe that it effectively destroyed the guerrillas to an extent that they were unable to exploit the popular unrest that the process generated. The one clear exception to this pattern of extreme repression was Venezuela between 1962 and 1965, where a democratic govern-ment's approach was more reminiscent of Western methods. Thus the astute Betancourt not only held his armed forces in check but also displayed exaggerated respect for due legal process. He believed firmly in minimum force and only acted to suspend civil rights temporarily when he was convinced that the majority of the population supported him. Nevertheless, the use of minimum force did not rule out maximum displays of force in terms of the presence of heavily-armed troops dominating the rooftops and crossroads of Caracas, thus both exhibiting resolve and reassuring the population, while deterring the guerrillas.

The guerrillas in Venezuela, although probably closer to

success than most other groups in Latin America in the 1960s or 1970s, ultimately failed to win enduring popular support and it must be said that the frequent success of counter-insurgency in the continent has derived as much from the inadequacies of guerrilla theory as from the efficiency of the Security Forces. In Venezuela, Betancourt's insistence on holding presidential elections in December 1963 in which 90 per cent of the eligible population voted, despite guerrilla attempts to enforce a boycott, proved a triumphant vindication of the Security Forces. In Uruguay, too, the Tupamaros' brief flirtation with conventional politics brought them little success at the polls in 1971.

'Foco' in the early 1960s was flawed by the assumption that the situation in Cuba in the late 1950s was unexceptional and that conditions favourable to the development of rural insurgency existed as a matter of course throughout Latin America. Far from providing the 'small motor of revolutionary dissolution', the 'foco' bands proved little more than aliens in a hostile environment. Rural insurgency thus lasted for the most part only as long as it took for the armed forces to reach the area of guerrilla activity. Urban guerrilla theory remains something of a contradiction in terms, the cities often proving, as Castro once remarked, 'the graveyard' of revolutionaries. Urban insurgency did little to shake the control of the armed forces where they already governed, and although it destroyed Uruguayan democracy, it did not succeed in supplanting government control. Even in a democratic system such as Venezuela, where the danger from terrorism was the most acute, insurgency only succeeded in strengthening the state. Thus there has been little success for guerrillas of any description in Central and South America since 1945 beyond Cuba and Nicaragua. Insurgency of course persists, guerrillas tending at present to have rediscovered the virtues of the countryside, as in El Salvador and also Peru, where the Maoist *Sendero Luminoso* group has been active since 1980, but it remains to be seen whether it will succeed. The price of that failure by the guerrilla movements of Latin America has, however, resulted in a general suppression of human rights throughout the continent. Of this, the outcome of insurgency in Uruguay is a clear example.

Amidst the other states of Latin America in the 1960s, where violence of one kind or another was rife, Uruguay stood out starkly as a distinct exception. Until the 1960s it had been a model of peaceful political and social development. It is a small country of approximately three million inhabitants. Most of the population are of European stock and almost all (some 80 per cent) live in the towns, mainly in the capital, Montevideo. The open countryside is very thinly populated and highly developed into large sheep and cattle ranches, the products of which form the bases of the country's economy.

This socio-economic pattern was developed in the last quarter of the nineteenth century, when a large influx of Spanish and Italian immigrants boosted the nation's workforce and also expanded the strong middle-class section of the population. Between 1903 and 1915 the government of Uruguay was in the hands of men of considerable political skill and reforming zeal, and it was from this time that the basis of political freedom was established, together with a remarkable degree of government-sponsored welfare which included free medical care, education and state pensions. In the first half of the twentieth century, the Uruguayan economy prospered and with it a democratic government and degree of civil liberty. The Army and Police were of insignificant proportions in a state which seemed removed from threats to its stability both from within and without.

Then, in the aftermath of the Korean War (1950–3), the cycle of economic development and almost universal prosperity came to an end as world demand for wool fell and prices slumped. From 1954, in a steady downwards economic spiral, Uruguay's gross national product (GNP) actually declined. The twin problems of unemployment and inflation escalated, with an annual inflation rate rising in the 1960s to as much as 135 per cent. In Montevideo, where over half the population lived, living standards, particularly among the manual working classes, declined dramatically. In the 1960s this sparked off a succession of strikes and lockouts which created, almost for the first time in Uruguay, an extremely high level of tension. In more prosperous times the proportion of workers employed in government service had increased considerably and in the 1960s it was estimated that a fifth of Uruguay's working population was ensconced in an ever-more inefficient and frequently corrupt bureaucracy. The decline in living standards and the ubiquitous corruption within government departments seem to have provided the motivation for the first really effective insurgency organisation that Uruguay had known — the Tupamaros.

The group, which took its popular name from Tupac Amaru, the last surviving member of the Inca royal family (executed by the Spanish in 1571), was formally known as the *Movimiento de Liberación Nacional* (MLN) and its membership was almost entirely made up of young men and women of middle-class background.[10] It was formed in 1963 and from the outset it based its political activities and later its guerrilla/terrorist campaign firmly on the urban area of Montevideo, rejecting the rural scenario. In the mid-1960s the gathering tension led to ever-more acts of violence: labour unrest was such that strikes abounded and the resultant confrontations with the forces of law and order were exacerbated by the politicisation of the student population. At first the MLN contented itself with organising a series of bank raids in an attempt to redistribute wealth in a 'Robin Hood' fashion and in so doing won a considerable degree of popular support. As the crisis developed, however, the 'Tupas', as they came to be called, began more and more to associate themselves with the labour movement and students in their various protests.

It was not until 1966 that the first armed confrontations between the Tupamaros and the police occurred.[11] They were of a minor nature initially, but marked the beginning of an escalation towards major violence. In one case — the very first — the police gave chase to a stolen van which eventually crashed, killing one of the occupants. During the chase, the men in the van — later confirmed as Tupamaros — fired at the police vehicle. Then, on 27 December 1966, a police superintendent leading a raid on a house occupied by the Tupamaros was killed in the firing which broke out as his party attempted to storm the building. Although these incidents did not arouse grave public anxiety, it gradually became apparent to the police that the Tupamaros were building up their strength. Weapons were being acquired and a rash of minor incidents during 1967 made up a pattern of events firmly linked to the dissident organisation.

Throughout 1966 and 1967 the economy of Uruguay continued to decline and at the same time a number of serious cases of corruption in government were brought to light, largely through the efforts of the Tupamaros who, with a mixture of skill and inside knowledge, made public the nefarious dealings of certain public figures and national companies. In 1967 Jose Pacheco Areco became President of Uruguay and, faced with a developing crisis in economic and security terms, he acted instantly to curb the rising clamour of public criticism directed against the government. In December he decreed the closure of two influential left-wing newspapers and also banned a group of extreme left-wing political parties. In June 1968 a state of emergency was declared in response to an upsurge of student rioting and militant trade union activity; apart from a brief period in 1969, such emergency powers were to be retained by the government throughout the remainder of the crisis. Nor did they seem unjustified — in August 1968 the Tupamaros kidnapped the president of the State Telephone Company, a close friend of President Areco and a man recently involved in a scandalous incident for which he had been sentenced to only a brief term in prison before being released. His kidnap received widespread acclaim among the middle classes of Montevideo. A series of other Tupamaros operations, including further kidnappings, bank and casino raids and the seizure of radio stations from which to broadcast propaganda messages, persisted throughout 1968 and 1969.

Steps were taken during this period by the government to improve the resources and effectiveness of the police and the armed forces. The former continued to bear responsibility for dealing with the emergency, although they were reinforced by a large paramilitary force (some 20,000 men), well-armed and trained specifically to respond to acts of urban terrorism. A substantial number of senior police officers were sent to the United States to attend courses on various aspects of counter-insurgency technique.

But they could do little to prevent continued Tupamaros attacks, which gradually became more daring. In October 1969 a group of about 40 guerrillas assaulted the town of Pando, some 32 km (20 miles) outside Montevideo, seizing the police station and capturing a number of weapons. Three banks were broken into and a large amount of cash, some of it in the form of US dollars, was stolen. The Security Forces reacted swiftly — police reinforcements were flown to the town by Air Force helicopters and all roads out of Pando were blocked — but, despite the fact that a number of Tupamaros were killed and others captured, it was apparent that the guerrillas had scored a major propaganda victory.[12] This was reinforced in May 1970 when about 20 Tupamaros, some dressed as policemen, gained access to the

Navy Training Centre in Montevideo, took the staff prisoner and made off with a very heavy haul of weapons and ammunition. It was a highly professional raid which left the Security Forces looking weak and ineffective.

Meanwhile, as unrest continued among students and trade unionists, the Tupamaros maintained their pressure upon the government by means of kidnappings and raids on banks and police stations, acquiring cash and weapons which significantly enhanced their influence, although it did prove difficult to translate this into effective political gains. What the guerrillas appear to have been trying to achieve was a 'climate of collapse', within which the government would be forced either to step down voluntarily from power or to make genuine concessions to the dissidents. As part of this policy, designed to embarrass the government and to draw attention internationally to the problems of Uruguay, the Tupamaros carried out two major kidnaps of foreign nationals. In July 1970 they abducted Dan Mitrione, an American official on loan to the Uruguayan government and working in police headquarters in Montevideo. His body was discovered on 10 August after the government had refused to accede to demands for the release of Tupamaros prisoners in exchange. In fact, the police enjoyed unexpected success during their response to the kidnap; on 7 August, acting on information, they surrounded a hospital run by Montevideo University and captured a large group of guerrillas, including many of the top Tupamaros leaders. Five months later the British ambassador to Uruguay, Geoffrey Jackson, was kidnapped, although on this occasion the guerrillas made no demands concerning a release; indeed, over the next few months a number of Uruguayan politicians were abducted and then set free, perhaps in a deliberate move designed to show how powerless the government had become. Jackson was held until September 1971, being released in the aftermath of a spectacular prison break in which over 100 Tupamaros escaped.[13]

In November 1971 presidential elections were held in Uruguay and although Areco failed in his bid for re-election, one of his supporters, Juan Maria Bordaberry, did succeed, probably as a result of ballot-rigging. He immediately declared his intention of destroying the Tupamaros and on 15 April 1972 obtained congressional approval for the declaration of a state of internal war throughout the country. This allowed him to bring the Army fully into the conflict and over the next few months, as civil liberties virtually disappeared, a ruthless and unexpectedly efficient military campaign succeeded in breaking up the guerrilla organisation. The main hiding-place for kidnap victims was located and over 800 leading Tupamaros activists were arrested. By November 1972 the organisation, which had never managed to gain widely-based popular support, had been effectively neutralised.

Unfortunately, in the process democracy disappeared. Having brought an end to the armed conflict, the Army, convinced that it was the corruption and failure of civilian political processes that had caused the crisis, entered the political arena. In February 1973 the armed forces openly challenged the president's decision to appoint a new minister of defence and forced Bordaberry to accept a number of political demands formulated by the chiefs of staff. These included the formation of a National Security Council in which the service chiefs would hold paramount influence, and it was through this body, in June 1973, that Congress was dissolved and all remaining organs of official opposition to government policy gradually banned. By the end of the year Uruguay was, to all intents and purposes, a military dictatorship in which Bordaberry continued to rule, subject to the demands of the armed forces.

In superficial terms, there can be no doubt that the Tupamaros were defeated — occasional acts of terrorism still occur in Uruguay, but their effect is severely limited and not apparently tied to an organisation with either popular support or political cohesion — and that the Uruguayan experience must be described as a successful counter-insurgency campaign. But there is more to it than that, for the price of success was heavy, involving the destruction of civil liberties and of the democratic process. Any liberal government facing an insurgency threat must always be aware that it possesses, in its armed forces, the capability totally to crush a minority movement, but such a purely military response rarely if ever works, chiefly because of its lack of subtlety and failure to distinguish between guerrillas and the ordinary people. The results will inevitably be repressive and will drive a wedge between government and people which, although the guerrillas themselves will be in no position to exploit it, will destroy the liberal nature of the state. The fact that such a temptation exists and that the Uruguayans, in common with many of their neighbours, fell victim to it, renders a deeper analysis of the Tupamaros campaign essential.[14]

Until the early 1960s, the general relative prosperity of Uruguay had meant that there was little significant unrest among the population. In consequence the police force was small and possessed of only limited resources. Its role was to maintain law and order in a society where lawlessness was by and large limited to isolated criminal acts rather than politically-motivated activities. With seemingly little external threat and no thought for the need to provide support to the police, the Army too was small, badly equipped and, inevitably, poorly trained and organised. Neither the police nor the military leadership aspired to political influence; both organisations were totally subordinated to the will of a democratically-elected government, but ill-prepared to face an insurgent threat. In such circumstances, the Tupamaros were able to carry out their initial attacks against the government and commercial interests without any real fear of effective counteraction. Indeed, in the prevailing climate of economic crisis and declining living standards, which affected the Security Forces as much as the rest of society, it was not unknown for members of the Police or Army actually to show sympathy for the Tupamaros cause.

Throughout the early years of the insurgency, up to 1972, the Army played only a minor and indirect role in counter-insurgency operations, although the police did gradually acquire more and better resources and developed a higher level of efficiency. The armed forces were used to support the police by providing training facilities and logistic back-up — on more than one occasion, for example, military helicopters were placed at the disposal of the police and used to telling effect — but there appears to have been a deliberate government policy to hold the generals, who were obviously watching developments with growing concern, firmly in check. The emphasis, as in other democratic societies facing insurgent activities, was upon the police as the representatives of civil rather than military power.

Unfortunately, as the Security Forces suffered blow after blow to their morale following each example of Tupamaros success, it became apparent that the police on their own were incapable of effective action, even when they were joined by the 20,000-strong paramilitary force in 1968–9. At the same time, it became increasingly difficult for the political command structure to maintain a firm grip on the military chain of command. Typically, if Security Forces acting in support of government policy meet with failure and are denied the freedom of action they deem necessary to rectify the situation, there will always be the risk of a collapse of authority within their hierarchy. This began to happen in Uruguay in 1971, when the military chiefs set up a joint service committee of senior officers tasked to study political and economic questions with a view to urging solutions to the crisis on the government, and when, in April 1972, Bordaberry at last invited the armed services to wage war against the Tupamaros, he probably only just pre-empted the military taking the law into their own hands. Under pressure from both the guerrillas and the armed forces, the president was choosing what appeared to be the lesser of two evils.

The result was that the military rolled into action against the Tupamaros and other dissident organisations without having to worry too much about close political control, and this was shown from the outset as force was used with scant regard for the liberal principles of a democratic state. Acting frequently on flimsy information, the Army cordoned off suspected areas and conducted heavy-handed searches. Large-scale arrests were made which inevitably involved innocent members of the public and subsequent interrogations were harsh, often to the point of torture. Because the Tupamaros managed to organise major breakouts from civilian prisons, most of the detainees were transferred to military compounds, where conditions were even worse and security much tighter. Within six months the Tupamaros organisation had virtually ceased to exist, but in the process the military grip on the state had tightened to the extent that a return to liberal democracy was impossible.

Thus, throughout their brief campaign, the armed forces relied on their overwhelming military power to subjugate the population. They had little or no previous experience of keeping the peace by acting in support of the civil authority and there is little evidence of a body of doctrine which guided their actions, based on police or military experience in their own or other countries. It was a purely military response, exploiting the power of the armed forces, and although in the short term the threat from the Tupamaros was countered, the long-term damage to the state was severe and large sectors of the population alienated from the forces of law and order.

The lack of previous experience in the subtleties of counter-insurgency was shown most markedly in the area of intelligence. In common with most urban guerrilla groups, the Tupamaros endeavoured to organise themselves on the basis of distinct and separate 'cells', hoping thereby to reduce the damage that one defection or discovery might do. However, their organisation was at no time particularly large and it never managed to win for itself mass popular support — as late as 1972, on the eve of the military campaign, it was estimated that there were fewer than 3000 active Tupamaros members — and these factors could have been exploited by Security Forces wishing to isolate the guerrillas from the bulk of the population. Indeed, as the 'Robin Hood' image of the Tupamaros waned and the public began to turn in increasing numbers against the organisation, the Security Forces even received valuable information about their enemy, but at no time could it be said that the military evolved a co-ordinated or effective infrastructure, dedicated to the gathering and collation of intelligence. There were very few examples of small carefully-executed operations achieving the degree of success that would have indicated a flow of precise and reliable information effectively analysed; the hideouts of the Tupamaros' kidnap victims were blundered on following the flooding of large areas with Security Forces, rather than scientifically isolated from a basis of collated information; the overall picture of the guerrilla organisation was never clear, with no attempt to attract defectors to the government side and very little information resulting from captured Tupamaros members, many of whom knew little beyond the confines of their own 'cell'. The result was a blanket response to the threat rather than a careful concentration upon the minority within the state who were causing trouble.

Inevitably, as the crisis developed through the 1960s and into the 1970s, the Uruguayan armed forces increased in size and effectiveness, acquiring many new weapons under the pretext of preparing to defend the state. Between 1968 and 1973 their share of the annual national budget rose from 13.9 per cent to 26.2 per cent; in 1963 it had been as little as one per cent. Some appreciation of the dramatic change involved may come from the realisation that whereas in 1969 some 990 million *pesos* were spent on the armed forces, by 1972 this had risen to 43,964 million — a truly staggering increase even allowing for a fall in the international value of the *peso* as Uruguay's economy deteriorated. Most of the arms, vehicles and equipment purchased from this enlarged budget came from the United States, ever anxious to prevent destabilisation in its Latin American sphere of influence, but the end result was a military force which enjoyed more power than was needed to subjugate a largely harmless population. Small wonder, therefore, that once the Tupamaros had been defeated in the summer of 1972 and law and order restored, military influence did not decline. Despite the long tradition in Uruguay of military non-interference in the politics of the state, the Army commanders, flushed with the degree of success they had achieved and wary of the apparent weakness of civilian rule, refused to return meekly to barracks. Having gained increased influence over the government during the period of the internal war, the generals continued to apply pressure on the president to ensure that the drastic and painful measures considered necessary to restore the country's fortunes were in fact carried out. By June 1973 all left-wing political activity had been eliminated and the Legislature dissolved, leaving Bordaberry as little more than a figure-head in a state devoid of democracy.

The Uruguayan example stands as typical of that strand of counter-insurgency which depends on overwhelming force and repression to succeed. There is little room for subtlety in such an approach, with its implied belief that the whole of the population is in some way guilty of supporting the guerrillas, and as a result there is little or no scope for policies which ensure continued popular support for the government and Security Forces. Thus, although the overt threat of insurgent activity may be quickly countered, the lack of a hearts and minds policy can and often does produce a different kind of crisis, in which the accepted forms of government are undermined and the principles of liberal democracy gradually destroyed. In the end, having chosen a drastic course in its efforts to survive guerrilla pressure, the government may well open the way to a different form of dictatorship based upon the use of force.

Notes

1. G. Kennedy, *The Military in the Third World* (Pall Mall, London, 1974), pp. 337–9.

2. R. Taber, *The War of the Flea* (Paladin, London, 1970), p. 39.

3. Carlos Marighela, *For the Liberation of Brazil* (Penguin Books Ltd., Harmondsworth, 1971), p. 46.

4. J. Kohl and J. Litt, *Urban Guerrilla Warfare in Latin America* (MIT, Cambridge, Mass., 1974), pp. 11–12.

5. R. Gott, *Guerrilla Movements in Latin America* (Nelson, London, 1970), p. 361.

6. D.S. Blaufarb, *The Counterinsurgency Era: US Doctrine and Performance* (Free Press, London and New York, 1977), pp. 279–88; Kohl and Litt, *Urban Guerrilla Warfare in Latin America*, pp. 8–14; R. Moss, *Urban Guerrillas* (Temple Smith, London, 1972), p. 157. See also the early view of US-inspired 'civic action' in W.F. Barber and C.N. Ronning, *Internal Security and Military Power: Counterinsurgency and Civic Action in Latin America* (Ohio State University Press, 1966).

7. D. James (ed.), *The Complete Bolivian Diaries of Che Guevara and Other Captured Documents* (Stein and Day, London, 1968), p. 23. For Peru, see L.M. Vega, *Guerrillas in Latin America* (Pall Mall, London, 1969), pp. 84–7.

8. V. Collazo-Davilia, 'The Guatemalan Insurrection' in B.E. O'Neill, W.R. Heaton and D.J. Alberts (eds.), *Insurgency in the Modern World* (Westview, Boulder, Colorado, 1980), pp. 109–36.

9. R. Gillespie, *Soldiers of Peron* (Oxford University Press, London, 1982), p. 247.

10. The social origins of the Tupamaros are covered in Moss, *Urban Guerrillas*, pp. 211–12; E. Halperin, *Terrorism in Latin America* (Washington Papers 33, Sage, London, 1973), pp. 37–48; and A.C. Porzecanski, *Uruguay's Tupamaros: The Urban Guerrilla* (Praeger, New York, 1973), pp. 28–32.

11. A detailed chronology of Tupamaros' activities can be found in Kohl and Litt, *Urban Guerrilla Warfare in Latin America*, pp. 196–226.

12. A detailed narrative of the Pando raid can also be found in Kohl and Litt, pp. 237–59, being reproduced from M.E. Gilio, *The Tupamaros* (Secker and Warburg, London, 1972).

13. Geoffrey Jackson, *People's Prison* (Faber and Faber, London, 1973).

14. A useful account of government counter-insurgency is to be found in J.A. Miller, 'Urban Terrorism in Uruguay: The Tupamaros' in O'Neill, Heaton and Alberts (eds.), *Insurgency in the Modern World*, pp. 137–88, especially pp. 164–74.

References

A. *General Works*

Banochea, R.L. and San Martin, M. *The Cuban Insurrection, 1952–59* (Transaction, New Brunswick, 1974)

Barber, W.F. and Ronning, C.N. *International Security and Military Power: Counter-insurgency and Civic Action in Latin America* (Ohio State University Press, 1966)

Blaufarb, D.S. *The Counterinsurgency Era: US Doctrine and Performance* (Free Press, London and New York, 1977)

Clutterbuck, R. *Protest and the Urban Guerrilla* (Cassell, London, 1973)

Debray, R. *Strategy for Revolution* (Penguin Books Ltd., Harmondsworth, 1973)

Debray, R. *Che's Guerrilla War* (Penguin Books Ltd., Harmondsworth, 1975)

Debray, R. *The Revolution on Trial* (Penguin Books Ltd., Harmondsworth, 1978)

Dunkerley, J. *The Long War: Dictatorship and Revolution in El Salvador*

(Junction, London, 1982)

Gillespie, R. *Soldiers of Peron* (Oxford University Press, London, 1982)

Guevara, E. *Bolivian Diary* (Cape, London, 1968)

Guevara, E. *Reminiscences of the Cuban Revolutionary War* (Allen and Unwin, London, 1968)

Guevara, E. *Episodes of the Revolutionary War* (International, New York, 1968)

Gott, R. *Guerrilla Movements in Latin America* (Nelson, London, 1970), revised as *Rural Guerrillas in Latin America* (Penguin Books Ltd., Harmondsworth, 1973)

Halperin, E. *Terrorism in Latin America* (Washington Papers 33, Sage, London, 1973)

James, D. (ed.) *The Complete Bolivian Diaries of Che Guevara and Other Captured Documents* (Stein and Day, London, 1968)

Kohl, J. and Litt, J. *Urban Guerrilla Warfare in Latin America* (MIT, Cambridge, Mass., 1974)

Marighela, C. *For the Liberation of Brazil* (Penguin Books Ltd., Harmondsworth, 1971)

Maullin, R. *Soldiers, Guerrillas and Politics in Colombia* (Rand, Santa Monica, California, 1973)

Moss, R. *Urban Guerrillas in Latin America* (Conflict Studies, London, Paper no. 8, 1970)

Moss, R. *Urban Guerrilla Warfare* (International Institute for Strategic Studies, London, Adelphi Paper no. 79, 1971) (*NB* This paper contains a transcript of Marighela's 'Minimanual of Urban Guerrilla Warfare')

Moss, R. *Urban Guerrillas* (Temple Smith, London, 1972)

Russell, C., Miller, J. and Hilder, R. 'The Urban Guerrilla in Latin America: A Select Bibliography', *Latin American Research Review*, 9/1, 1974, 37–79

Taber, R. *The War of the Flea* (Paladin, London, 1970)

Thomas, H. *Cuba or the Pursuit of Freedom* (Eyre and Spottiswoode, London, 1971)

Vega, L.M. *Guerrillas in Latin America* (Pall Mall, London, 1969)

B. *The Tupamaros*

Hughes, D.C. *The Philosophy of the Urban Guerrilla: The Revolutionary Writings of Abraham Guillen* (William Morrow, New York, 1973)

Gilio, M.E. *The Tupamaros* (Secker and Warburg, London, 1972)

Labrouso, A. *The Tupamaros* (Penguin Books Ltd., Harmondsworth, 1973)

Miller, J.A. 'Urban Terrorism in Uruguay: The Tupamaros' in B.E.O'Neill, W.R. Heaton, D.J. Alberts (ed.), *Insurgency in the Modern World* (Westview, Boulder, Colorado, 1980), pp. 137–88

Moss, R. *Uruguay: Terrorism versus Democracy* (Institute of Conflict Studies no. 14, London, 1971)

Porzecanski, A.C. *Uruguay's Tupamaros: The Urban Guerrilla* (Praeger, New York, 1973)

5 THE PORTUGUESE ARMY: THE CAMPAIGN IN MOZAMBIQUE, 1964–1974

Ian F.W. Beckett

Portugal was the first European colonial power to arrive in Africa, a presence being established in each of the three principal colonies of Portuguese Guinea, Angola and Mozambique by the end of the fifteenth century. Coastal occupation, however, did not constitute control and it was only gradually that the Portuguese were able to assert their authority over the interiors. There was a rash of colonial campaigns in the late nineteenth and early twentieth centuries, culminating in the suppression of the last significant rising in Guinea as late as 1936. When a nationalist revolt erupted in Angola in February 1961, therefore, the Army had seen no major action since the First World War, Portugal having remained neutral during the Second.

Almost simultaneously, the Army suffered a humiliating defeat at the hands of the Indian Army, when Portuguese Prime Minister Antonio de Oliveira Salazar refused to recognise the realities of Indian independence and provoked an invasion of the Portuguese colony of Goa and its enclaves on the west coast of the sub-continent. Ordered to fight to the last man, the outnumbered and ill-equipped garrison capitulated after only two days on 18 December 1961. The mutual recriminations which ensued between Army and politicians were to have some relevance to the end of the Army's African campaigns some 13 years later, as the outbreak of revolt in Angola was followed by the beginning of hostilities in Guinea in January 1963 and in Mozambique in September 1964. In many respects it was a considerable achievement for the Army of arguably the poorest and least developed country in Western Europe to fight three wars simultaneously for such a prolonged period without suffering military defeat. But the strains which resulted contributed markedly to a breakdown of civil-military relations and a military coup in April 1974 which would destroy the fabric of the Salazarist state and its 'historic' mission in Africa.

As in most counter-insurgency campaigns since 1945, the

Map 5.1: Mozambique 1974

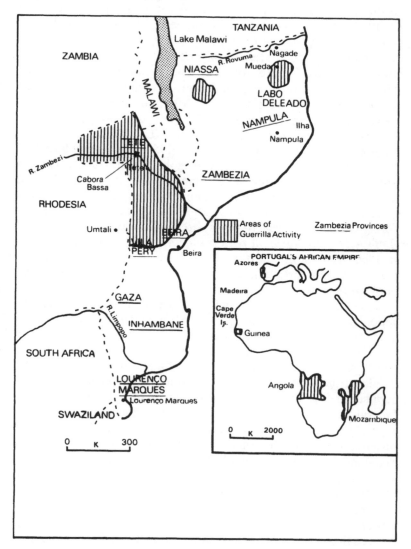

situations pertaining in the colonies were complex and none more so than in Angola, where three rival nationalist groupings were eventually fighting against the Portuguese. That known as MPLA (*Movimento Popular de Libertação de Angola*) had been founded in Luanda in 1956, its leadership dominated by *mesticos* (half-castes) such as Viriato da Cruz and *assimilados* (assimilated Africans) such as the doctor and poet, Agostinho Neto. The rank and file, however, were primarily Mbundu tribesmen from the areas surrounding the capital. By contrast UPA (*União das Populações de Angola*), later renamed FNLA (*Frente Nacional de Libertação de Angola*) and founded in 1954, was largely based on the Bakongo tribe of northern Angola. Their leader, Holden Roberto, was the nephew of one of the tribe's traditional kingmakers. As in the other colonies, there were subsequent splits within the two main nationalist groupings, Roberto's 'foreign minister', Jonas Savimbi, breaking away in 1964 to found UNITA (*União Nacional para a Independencia Total de Angola*) in 1966. Operating mostly in southern Angola, UNITA was heavily dependent upon the support of the Ovimbundu and Chokwe tribes.

In Guinea there were similar divisions between Portugal's opponents, the main group — PAIGC (*Partido Africano da Independência de Guiné e Cabo Verde*) — being dominated by Cape Verdean *mesticos* such as the Portuguese-trained agronomist, Amilcar Cabral, but with the rank and file drawn from the Balante tribe who comprised only some 31 per cent of the total population. Briefly a rival group known as FLING (*Frente para a Libertação e Independência de Guiné*) also flourished, displaying a particular animosity for Cape Verdeans, while PAIGC itself was not free from internal rivalries between mainlanders and islanders. In Mozambique FRELIMO (*Frente de Libertação de Moçambique*) was heavily reliant upon the support of the Makonde tribe of the Mueda plateau who represented barely two per cent of the population as a whole, although FRELIMO subsequently drew support from the Nyanja of the Niassa region. As a coalition of earlier groups founded in 1962, FRELIMO also had its internal divisions, between Makonde traditionalists and FRELIMO ideologists and there were a number of dissident groups such as COREMO (*Comite Revoluçionario de Mocambique*) which operated for a time in the west. There was a further split within FRELIMO between the political and military wings of the movement which led, after the assassination of its first leader, Dr Edouardo Mondlane, to the dominance of the military wing represented by Samora Machel.

In each of the colonies, however, the Portuguese could themselves derive support from other tribes, particularly the Moslem Fula and Mandinka in Guinea and the largely Moslem Macua in Mozambique, who were traditional enemies of the Makonde. Such demographic factors were also of significance to both sides in that some tribes were spread across international frontiers, as in the case of the Bakongo living astride the Angola-Zaire frontier and the Makonde living on both sides of the Mozambique/Tanzania frontier. If this largely benefited the guerrillas, it could on occasions work to the advantage of the Portuguese. FRELIMO found it difficult to penetrate south of the Cabo Delgado and Niassa regions because of the concentration of the hostile Macua across central Mozambique. Similarly, MPLA was unable to operate out of Zaire due to FNLA's influence there and was forced, first, to attempt to infiltrate columns into its own Mbundu territory from Zambia and, second, to mount operations in the virtually uninhabited regions of eastern Angola.

The existence of such international frontiers, providing readily-accessible guerrilla sanctuaries, represented a considerable difficulty for the Portuguese, who were largely constrained by political considerations from striking across them or indulging in 'hot pursuit'. With the exception of UNITA, which was expelled from Zambia for sabotaging the economically-vital Benguela railway running through Portuguese territory, all the nationalist guerrilla groups found external refuges. FNLA operated out of Zaire, MPLA out of the Congo Republic and Zambia, PAIGC out of the Republic of Guinea and Senegal and FRELIMO out of Tanzania, Zambia and, latterly, Malawi. There were occasional claims by the guerrillas' hosts that Portuguese forces had crossed the frontiers. Senegal, for example, complained of incursions in September 1967, November and December 1969, May 1972 and May 1973, while accusing the Portuguese of laying mines inside Senegalese territory in July 1971. There were Zambian accusations of Portuguese incursions in February and November 1968 and in June and July 1969. Zaire claimed an incursion in March 1969 and Tanzania shot down a Portuguese Harvard T6 converted trainer in April 1972. The relatively small number of such claims, however, is an indication of Portuguese reluctance to incur additional international criticism, although they were certainly implicated in an abortive seaborne invasion by exiles of the Republic of Guinea on 22 November 1970.

But if direct military action was seldom required, this did not preclude the use of an espionage network in Zambia and Tanzania and the exploitation of divisions within the nationalist groups. Thus the Portuguese secret police, PIDE (International Police for Defence of the State; later DGS — Directorate General of Security), may well have been involved in the internal conspiracies which resulted in the assassination of Mondlane in February 1969 and of Cabral in January 1973. Some leverage could also be extracted from the dependence of states such as Zambia and Malawi on the Portuguese rail system, the Portuguese briefly imposing an economic blockade on Zambia in March 1971 after the abduction of some Portuguese farm workers by COREMO.

There was, however, little that the Portuguese could do to prevent the training undertaken by the guerrillas in their hosts' territory or the assistance offered to the guerrillas by states further afield. The first 250 FRELIMO guerrillas were trained in Algeria, as were the first 300 MPLA guerrillas, while the latter also enjoyed facilities in Bulgaria, Czechoslovakia and the Soviet Union. PAIGC and FRELIMO similarly received the benefit of Soviet assistance, while the Chinese supported FRELIMO, FNLA and especially UNITA. FNLA even derived tacit support from the United States and FRELIMO had many American sympathisers while Mondlane, who was married to a white American, was still alive. Such support could be more than just material, PAIGC incursions sometimes being covered by Guinean artillery fire. The Portuguese captured a Cuban inside Guinea in 1969 and killed a further four there in 1970, while Nigerian pilots flew MiG-17 reconnaissance flights over the colony from 1971 onwards. After the abortive invasion of the Republic of Guinea, both Nigeria and Egypt offered troops to the Guinean government and a Soviet naval force was deployed off the coast to prevent repetition. The Portuguese also had to contend with the attentions of a wide range of nationalist fellow-travellers and propagandists. Not unexpectedly, the United Nations was a constant and vociferous critic of Portuguese policy. A UN delegation allegedly toured 'liberated zones' in Guinea in March

1972, although there is considerable doubt that they actually entered the colony. When PAIGC proclaimed an 'independent' state in September 1973, it was immediately recognised by 70 countries, welcomed in the UN General Assembly and admitted to the Organisation of African Unity (OAU). Similarly, although the Portuguese secured a visit by Pope Paul VI to the Fatima shrine in metropolitan Portugal in 1967, the guerrillas scored equal success when Cabral, Neto and Marcelino dos Santos of FRELIMO were granted an audience with the Pope in Rome in July 1970.

Portugal derived some support from membership of the North Atlantic Treaty Organisation (NATO), having been a founder member in 1949. Although the United States refused to allow the Portuguese to deploy the F-86 Sabrejet in the colonies from 1967 onwards, all Portugal's military equipment was supplied by NATO allies, such as the standard G3 German carbine with which the infantry were armed, the French Panhard and British Daimler armoured cars, the French Alouette helicopters and the Italian Fiat G-92 fighter that was the mainstay of Portuguese aerial operations in Africa. It can be noted, however, that most of the aircraft were elderly by NATO standards, 17 aircraft types being in use in 1973. The Portuguese argued that they were essentially fighting the West's battle against communism and that the continued possession of the Cape Verde Islands in particular was of great strategic importance to the West's control of the Atlantic sea routes. Within Africa itself the Portuguese could also count on the support of other white powers. It was frequently alleged that South African forces were active in Angola, particularly in defence of the important Cunene hydroelectric project in the south and Portuguese forces in Mozambique did in fact intercept a joint COREMO/PAC (Pan Africanist Congress) guerrilla force *en route* for South Africa in June 1968. Similarly, although nominally complying with international sanctions against Rhodesia, the Portuguese had contacts with the Rhodesian Security Forces at least from 1969; an unofficial 'Council of Three', drawing representatives from Rhodesia, Portugal and South Africa, being in existence from February 1971. In April 1973 Portugal's commander-in-chief in Mozambique, Kaulza de Arriaga, disclosed a 'gentleman's agreement' enabling Rhodesian forces to operate up to 100 km (62 miles) inside Portuguese territory.

The intractable determination of Salazar and his successor as Prime Minister, Marcello Caetano, to remain in Africa went some considerable way to offsetting the alignment of forces against Portugal. There was undoubtedly a genuine belief in a 'civilising mission' in Africa. Above all, there was a realisation that the continued possession of colonies conferred an international standing Portugal could not otherwise have attained, a concept neatly expressed by Caetano in the 1930s: 'Without Africa we would be a small nation; with Africa we are a big power.'[1] There was thus no lack of resolve in deploying sufficient numbers of troops to safeguard Portuguese territories. By 1967 some 75 per cent of the Metropolitan Army was overseas, in contrast to the minimal white Portuguese presence in the colonial garrisons of the past. Troop levels in Portuguese Guinea were increased from approximately 1000 in 1961 to over 30,000 by 1967; in Angola from 3000 in 1961 to over 60,000 by 1971; and in Mozambique from 16,000 in 1964 to over 60,000 by 1974. With over 150,000 men in Africa by 1970, the Portuguese deployment represented a troop level, in proportion to the Portuguese population as a whole, five times greater than that of the United States in Vietnam in the same year. Portugal regularly called up a greater proportion of those eligible for military service than any other country with the exceptions of Israel and the two Vietnams.[2]

Until his stroke in September 1968, Salazar effectively controlled overall colonial strategy in so far as co-ordination existed. Thereafter Caetano revived the Supreme Defence Council and gave the Minister of Defence additional authority by subordinating individual service ministers to him. Within the colonies themselves, Governors-General theoretically presided over the administration, but their powers were restricted to routine matters and the maintenance of law and order, while economic matters were largely handled by external commercial concerns such as CUF (*Companhia União Fabril*). Governors-General had no staffs to assist them until July 1969, while the Commanders-in-Chief in the colonies not only tended to lack overall command of the forces available but had no real staffs beyond small military cabinets. Frequently the powers of Governors-General and Commanders-in-Chief were vested in one individual or the Governor-General was a soldier. This should have led to a greater degree of co-ordination, yet even

Arnaldo Schultz, who held both posts simultaneously in Guinea between May 1964 and May 1968, effectively lacked full powers over both civil and military administrations through Lisbon's interference. It was left to his successor, Antonio de Spinola, to assume full control largely through dictating his own terms of appointment. Significantly such powers were exercised after Salazar's departure from office.

Joint military and civil committees did exist for operations, intelligence and psychological warfare but seldom worked well at lower level through the shortage of civilian administrators, with the result that many areas were fully taken over by the military under emergency powers. Where they did work well, the joint intelligence committees yielded results, but the Portuguese Army lacked an intelligence corps of its own and was hampered by the frequent turnover in staff. Much therefore depended upon the work of the civilian agencies and in Mozambique in particular there was considerable friction between Army and PIDE/DGS. Such problems were also compounded by the difficulty of infiltrating the guerrilla groups, with the Portuguese largely dependent upon information derived from defectors or captured insurgents or documents. It was not always easy, however, to turn background information of this kind into contact intelligence.

By 1970 a command system had evolved for military operations by which each colony was divided into a number of territorial or theatre commands — there were five such commands in Angola and three in Mozambique — with joint military headquarters in each theatre. At theatre level, elite units were made available for an intervention role in 'reduction' operations such as clearing areas or protecting urban centres from sabotage. Below theatre level, tactical responsibility was subordinated into sector commands further subdivided into battalion and company areas, although intervention forces were also available at local level for immediate reinforcement, convoy escort or similar mobile roles. As indicated already, this system only gradually developed as troop levels were raised and initially the Portuguese were thrown onto the defensive in both Angola and Guinea. In the former, none of the 3000 garrison in 1961 was in the northern areas of Uige where the revolt broke out, while there were only two Portuguese infantry companies in the whole of Guinea in 1963, the Portuguese admitting to losing control of some 15 per cent of the territory by the end of that year. Much of the immediate response to insurgency in Angola, therefore, was in the form of the available airpower to try and contain the revolt. Some accounts claim that 50,000 Africans were killed between August and September 1961 by the ferocious Portuguese counter-blows. FNLA were, however, very poorly armed by contrast to the better-trained PAIGC guerrillas in Guinea and, after a disastrous attempt to drive the guerrillas from Como Island in February 1964, the Portuguese surrendered much of the initiative to the PAIGC by withdrawing into outposts defended, in the manner of American firebases in Vietnam, by elderly 140 mm howitzers. Similarly, although having considerably more time to prepare, the initial Portuguese response in Mozambique was to hold the Rovuma River line separating the colony from Tanzania. To a certain extent, of course, the Portuguese emphasis on containing insurgency within remote areas was understandable and sensible and it remained a feature of their strategic approach until the end of their African campaigns.

More positive action was required to destroy the insurgents' capacity to infiltrate Portuguese territory and this became possible as troop levels increased and newer equipment was received. The arrival of the helicopter was the most significant factor in this enhanced strategy. Helicopters were first deployed in Guinea in 1967, some twelve machines eventually becoming available there. In Angola their arrival coincided with the opening of MPLA's 'eastern front' in the Moxico and Bié regions in 1966–7, some 60 Alouette helicopters being deployed by 1971. Capable of carrying five men and of mounting 20 mm grenade launcher or machine guns, the Alouette gave the Portuguese the ability to drop men where required to reinforce ground patrols and to cut off guerrilla escape routes. They were thus of major value in the large-scale dry season offensives (May to September) mounted in eastern regions of Angola in 1966, 1968 and 1972. In addition Portuguese light bombers could be utilised to seal off guerrilla supply routes across the frontiers, provide an immediate reaction force and attack villages where crops were being cultivated by the guerrillas.

The co-ordination of aircraft, ground patrol and helicopter, coupled with the initial unfamiliarity of the guerrilla with the latter, proved particularly potent. The Portuguese claimed, for example, that Operation 'Attila' in 1972 eliminated half the guerrillas operating in eastern Angola. Certainly Portuguese success and internal splits within MPLA had forced the guerrillas out of Zambia by 1974 and back to their original bases in the Congo Republic. However, although such large drives could prove successful where the terrain and the season were favourable, there was a growing emphasis on small-unit operations, since larger efforts only tended to force the guerrillas to operate elsewhere. Smaller continuous actions offered the possibility of more thoroughly disrupting guerrilla activities. In the heavily forested Dembos region of Angola, for example, one observer of Portuguese operations found that 30-man patrols for three to five days were the norm in 1968–9, while the same observer described similar 30-man patrols following predetermined courses to avoid mines and airstrikes in Guinea in 1971.[3]

Ambitious sweeps were therefore becoming a thing of the past by the late 1960s, although it has been suggested that there was a regrettable tendency for incoming Commanders-in-Chief to mount such large-scale operations at the beginning of their tenure.[4] For most of the period of the wars, however, the Portuguese had relatively little contact with insurgents whose main and most effective retaliatory weapon was the mine. Indeed, the war in Angola has been characterised as one of 'mines versus helicopters',[5] 50 per cent of Portuguese casualties in 1970 being attributed to mines. The Portuguese response was an extensive programme of road-tarring in each of the colonies, amounting to 8000 km (4970 miles) of tarred road in Angola by 1974 and an annual rate of 1400 km (870 miles) in Mozambique by 1972 which far exceeded that achieved by the British in twelve years in Malaya or by the Americans in six years in Vietnam. Rather less sophisticated, since many guerrilla mines were difficult to trace with mine detectors, was the troops' habitual use of *pica* or sharpened sticks to probe the ground ahead of patrols. Convoys were invariably preceded by Berliet trucks with sandbagged floors and tyres filled with water to absorb the blast.

This aspect of Portuguese operations is an indication that technology is not necessarily of paramount importance in counter-insurgency. Thus cavalry were also used efficiently, three squadrons being deployed in Angola from 1966 and further cavalry being under training in Mozambique by 1974. Cavalry could be utilised to protect the flanks of advancing troops in difficult terrain or, supplied by helicopter, could undertake their own long-duration patrols. It might be added that the horse was itself an effective shock absorber in the event of mine detonation. Similarly, although defoliants were widely used from 1970 onwards to clear roadsides of vegetation which might conceal ambushes, the lobbing of grenades into the verges ahead of troops evacuating vehicles was equally effective. The empty beer bottles slung on wires around Portuguese outposts in Guinea as an early warning device were just as successful as sophisticated sensors which, in any case, were not available.

The guerrillas, however, became more and more schooled in tactics and better armed year by year. To a large extent this only resulted in longer-range bombardment of Portuguese posts, particularly by heavy mortars and 122 mm rockets, but by 1973 the Portuguese monopoly of airpower was being seriously challenged for the first time. In Guinea, surface-to-air missiles (SAMs) appeared for the first time in March 1973, three aircraft being lost within two months and PAIGC claiming to have downed 21 aircraft by September. By March 1974 SAMs had also made their appearance in Mozambique. Measures could be introduced to combat the SAMs, such as helicopters flying below the 122-237 m (400-500 ft) height for which the missiles were armed or aircraft flying at maximum height or taking off and landing in tight circles. But there was no doubt that the slow-flying Alouettes and subsonic Fiats were vulnerable. Similarly, it was believed that Soviet-supplied tanks were also being used for training purposes by PAIGC guerrillas, although they were not to make a debut against the Portuguese.

While it can be argued that the Portuguese successfully adapted to the military requirements of counter-insurgency, they were less able to develop a convincing 'hearts and minds' policy to win support from the native population as a whole. One clear indication of the flawed socio-economic aspects of Portuguese counter-insurgency was the resettlement programme. It is generally recognised that any attempt to win the confidence of the indigenous population must be preceded by affording them adequate protection from guerrilla intimidation and by generally preventing guerrilla infiltration. On occasions this has been attempted by interposing a physical barrier between guerrilla and population in the manner of the Morice Line constructed by the French in Algeria or the Rhodesian *cordon sanitaire* of border minefields. More usually, since 1945, population and guerrillas have been separated by resettling the former in protected villages. Variously known as *senzalas do paz* or, more familiarly, as *aldeamentos*, Portuguese resettlement villages were first introduced into the Uige region of northern Angola in August 1961 to accommodate refugees returning from Zaire after the initial revolt had subsided. Up to 150 *aldeamentos* were established in the north, but the programme was greatly accelerated with their extension to the eastern regions of Angola in 1967 and to central and even southern Angola in 1968. In all it is estimated that a million people (or roughly 20 per cent of the native population) were resettled in Angola by 1974. In Guinea *aldeamentos* were first introduced by Schultz and, under Spinola, eventually embraced approximately 150,000 people, while a further million (or roughly 15 per cent of the native population) were resettled in Mozambique.

Resettlement was, however, largely a failure. Evidence from Angola indicates a division within the Portuguese authorities between those viewing resettlement primarily as a measure of population control and those regarding it as an opportunity to stimulate native development. Indeed it would appear that the extension of resettlement in the form of civilian-run *reordamentos* to central and southern Angola, where there was little immediate guerrilla threat, owed rather more to a desire to release land for further white settlement than to fears for the security of the native population. Resettlement was welcomed on occasions by those African tribal leaders who had reason to fear guerrilla intimidation, but more usually it was resented as disrupting traditional tribal society such as that of the semi-nomadic Herrero and Ovambo of southern Angola. The villages themselves were often badly sited without the advertised facilities. Areas available for cultivation were not only less productive than those from which the villages had been removed, but also subject to the depredations of wild animals at night. Rural agriculture generally appears to have suffered with, for example, manioc production in the Moxico region of Angola alone declining by 90 per cent in 1969. There is also evidence that inadequate sifting of the population led to continued penetration of the settlements by guerrillas or their supporters: in Mozambique it was claimed that a third of the food grown in *aldeamentos* was going straight to FRELIMO guerrillas.[6] At least resettlement effectively depopulated frontier areas and this, coupled with the 'maize war' by which the. Portuguese attacked guerrilla agriculture production and the natural root crops of the jungle areas with defoliants and herbicides, did have the effect of making infiltration much more difficult. What it did not do was to make winning hearts and minds any easier.

The 'social promotion' or psycho-social programme as it was often called, was dogged by lack of enthusiasm and a general lack of resources. There was certainly a greater preference for building roads than providing further facilities in *aldeamentos*, six times as much finance being spent on roads than on either health or education in Angola. The third Angolan development plan of 1968–73 devoted half its capital to road building since it was argued by one official that 'revolt starts where the road ends'.[7] It must also be remembered that Portugal itself was a backward country with low standards of literacy and development. Given those limitations, the Portuguese effort was not unimpressive. In Angola, for example, the number of schools increased from 3589 in 1967 to 5000 by 1972. The most comprehensive programme was that of Spinola in Guinea. As already indicated, Schultz had begun resettlement and instituted other programmes such as starting *cadmils* — co-operatives to dig wells, etc. Under the slogan of *Guiné Melhor* ('Better Guinea'), Spinola greatly accelerated the Portuguese programmes, building over 15,000 houses, 164 schools, 40 hospitals and 86 water points. Campaigns were waged against disease in both humans and cattle. Spinola claimed to be winning back 300 refugees a year from the Republic of Guinea and Senegal and large numbers of Senegalese regularly crossed the frontier to avail themselves of Portuguese medical facilities. Price controls were imposed in Guinea to restrain the cost of living for the native population and 'peoples' congresses' enabled them to air grievances. Overall the daily lot of the African probably did improve throughout Portuguese Africa, but not to the extent to which it did for white settlers and, of course, the Portuguese were unable to satisfy wider aspirations.

The initial revolt of 1961 had been followed by immediate reforms, including the abolition of compulsory cultivation of cash crops and the end of the system of assimilation; henceforth any African was theoretically equal to any European as a Portuguese citizen. In 1971 a new constitution granted greater autonomy to the overseas provinces within the Portuguese nation, this being enacted in 1972. Elections were held, for example, in Guinea in March 1973. How far the latter would have borne fruit in the long term is difficult to assess, but there is evidence that Portuguese 'counter-ideology' did have its successes, if only to judge by the number of Africans who fought for the Portuguese. Estimates vary considerably, but with a considerable expansion in the number of locally-raised forces from 1966 onwards, it would appear that 30–40 per cent of Portuguese forces in Angola, 50 per cent in Guinea and over 60 per cent in Mozambique were black by the end of the wars.[8] Some were conscripted, some were undoubtedly attracted by the pay and others by tribal animosities while, as a whole, the Portuguese employment of local forces was entirely pragmatic rather than ideological. The proportion of black troops was as high as 90 per cent in some elite units such as the *Grupos Especiais* (GE), airborne *Grupos Especiais de Paraquedist* (GEP), or Spinola's *Commandos Africanos*, while the DGS also recruited its own black intelligence-gathering units or *Flechas* ('Arrows'), some of whom were former guerrillas. Indeed Arriaga in Mozambique claimed a 90 per cent success rate in persuading captured guerrillas to turn against their former colleagues, the payment of cash rewards for weapons and the widespread distribution of surrender leaflets being common to all the colonies. Spinola even made a point of evacuating wounded guerrillas to hospitals before his own men. In addition there were black militias guarding *aldeamentos* although, in Angola, only about 10 per cent of the 30,000-strong militia was actually armed in 1974 and their main function appeared to be to compromise their members in the eyes of the guerrillas. In Guinea Spinola also armed the Moslem Fula with rifles, 15,000 natives being so armed by 1973, while Moslem support was also secured by such devices as arranging for tribal leaders to go on pilgrimages to Mecca. White settlers also served in local units.

The pragmatism of raising local forces is in itself an indication of how far armed forces may have to diverge from their former practices to meet the threat of insurgency. But, of course, counter-insurgency may require more fundamental changes in doctrine and training. In the case of the Portuguese Army, the conventional NATO organisation had to be completely abandoned for colonial operations, with the infantry battalion becoming the basic unit, with suitable fire support. Armour could not be deployed because of the difficult terrain, the sheer cost of maintenance and the vulnerability to mines or portable anti-tank weapons. Similarly, artillery could rarely be effectively used except for harassing fire on suspected guerrilla positions by night or in support of static bases. As a result, armoured and artillery units were forced to become infantry to the detriment of their conventional skills. All Portuguese troops destined for the colonies underwent a six-week training course in small-group tactics immediately after their basic training, to be followed by a four-week course of field exercises in the wooded areas of metropolitan Portugal before embarkation. In the same way, the Portuguese Air Force had to adapt to the requirements of counter-insurgency through close centralisation of command and control to capitalise upon the limited numbers of aircraft available in any one colony. Indeed commercial aircraft were regularly pressed into service for transportation while the Air Force concentrated on offensive air support. Only the Portuguese Navy retained its NATO role and organisation unchanged, although naval forces, particularly marines and naval fusiliers, were utilised in the colonies. In Guinea, large areas were either tidal or readily accessible to small rubber assault boats, while in Mozambique naval craft patrolled Lake Malawi and sixteen 50-ton patrol boats armed with 20 mm Oerlikons undertook anti-infiltration duties along the coast.

The wars had more than just a doctrinal impact upon the Portuguese armed forces and the Portuguese state. The cost in financial terms of maintaining the armed forces in Africa was enormous, the proportion of the national budget devoted to defence rising from approximately 25 per cent in 1960 to a peak of 42 per cent in 1968. This represented over six per cent of GNP, the military cost of the campaigns in real terms rising from 170.8 million *escudos* in 1967 to 4438.8 million *escudos* by 1973,[9] although it is possible that there were further hidden costs in such areas as transportation expenditure. Of the colonies, only Angola could go any way towards meeting such costs herself through the increasing economic potential of the Cabinda oilfields, the Diamang diamond mines, Cassinga iron-ore mines and the coffee and cotton plantations of the north. Angola thus provided 50 per cent of the cost of its defence in 1971, compared to only 20 per cent in Mozambique. In human terms, the cost was also great. The Portuguese admitted to a total of 3265 men killed in action and 3075 dead from other causes (such as the diseases common in tropical climes) in their campaigns between 1961 and 1974, but it is clear that these figures are understated and probably omit black troops altogether. A more realistic estimate is that of A.J. Venter, who calculated a loss of some 13,000 Portuguese dead by 1973, with some 65,000 non-fatal casualties from all causes between 1961 and 1973.[10]

There were, in addition, probably 80,000 civilian casualties and between 100,000 and 150,000 guerrilla casualties in the three colonies together. Certainly the strain on Portuguese manpower was such as to cause the age of conscription to be lowered to 18 years in 1967 and the term of conscript service extended from two to four years by addition of a compulsory two years overseas. However, not all those called actually served. As opposition to the wars grew in Portugal itself, so avoidance of the draft increased. It is generally accepted that 110,000 Portuguese declined to report for military service between 1961 and 1974 either through deliberate avoidance or absence abroad, since over a million Portuguese were emigrant workers by 1974. In the last call-up before the coup of April 1974 over 50 per cent did not appear.

Inevitably such strains in Portuguese society were reflected in the armed forces overseas. There was reputedly a problem with heavy drinking but there is no evidence of widespread drug abuse as occurred with American forces in Vietnam. Nor was desertion high. The Portuguese claimed in 1971 that the desertion rate had been no more than four or five per 1000 in the last four years. Troops in the colonies received a 25 per cent pay increase from 1971, the casualty evacuation record was good and attention was paid to important factors such as regular receipt of mail, accurate information and adequate food. Indeed, as most white Portuguese troops were peasants used to deprivation, army conditions were often better than those they had experienced at home. Nevertheless, there was a growing sense of isolation, as those remaining at home appeared to have little interest in the campaigns. Casualty figures were hidden in the back pages of the newspapers. The greatest problem appears to have been lack of motivation among many troops and there was certainly an element of bifurcation between the elite units such as para-troopers and commandos, who did the bulk of the fighting, and the great majority of conscripts who provided that 50 per cent of the Army in Guinea or 70 per cent of the Army in Angola employed almost exclusively on 'social promotion'. There were reports in Angola by 1970 that Portuguese ground patrols were increasingly black in composition or of only two or three white Portuguese with 20 to 30 black troops. In Guinea Portuguese troops were reported as taking a regular two-and-a-half hour siesta in the middle of the day, while the Rhodesians alleged that the Portuguese forces operating in the Tete region of Mozambique from 1972 onwards did so with the maximum amount of noise possible to ensure little contact with the guerrillas. Most of the atrocities alleged against the Portuguese forces, such as the disputed Wiriyamu affair of December 1972, which was significantly publicised just before Caetano's visit to London in July 1975, appear to have been associated with black troops such as the GE and GEP. All in all, however, there is some reason to doubt that there was as wide a collapse of morale as has been suggested by General Costa Gomes' remarks in Lourenço Marques in May 1974 that the Army had reached 'the limits of neuro-psychological exhaustion'.[11]

It must also be remembered that the coup of April 1974 resulted from a combination of purely domestic factors and more mundane grievances within the Army's officer corps which had only a peripheral relevance to the campaigns in Africa. The Portuguese officer corps had a tradition of political involvement. Indeed there were at least nine attempted coups against Salazar and three against Caetano. Although the great expansion of the officer corps in the 1960s and 1970s and its subsequent dilution in terms of social composition was undoubtedly due to the demands of war, the way had been paved by Salazar's waiving of fees for tuition at the military academy in 1958. As a result the officer corps became more lower middle class, more provincial and more democratic in outlook. It is to this process rather than any exposure of interrogating officers to the ideologies of captured guerrillas that the politicisation of the officer corps must be mainly attributed. Pay was poor by comparison with other European armies and promotion not only slow but subject to perceptions of political reliability. After Caetano devolved responsibility for promotion to the Supreme Defence Council, it became a matter of equally unsatisfactory strict seniority. There were divisions within the officer corps, most notably between line officers and those of the staff, Portugal being one of the few armies in the world to retain an antiquated separate staff corps.

There was also resentment on the part of many younger officers, forced to 'moonlight' to make ends meet, that leading general officers were able to take posts with large commercial concerns at considerable personal reward. Certainly constant overseas tours, the antagonism in the colonies between Army and white settler communities and the problem of motivating unwilling conscripts added to the sense of resentment, even though officers were generally held by white settlers to be fighting an 'air-conditioned war'[12] of some comfort.

The problem of motivation was exacerbated by the need to conscript large numbers of university graduates or *milicianos* as officers since the number of candidates for officer cadetships declined by an astonishing 300 per cent between 1961 and 1974. It was held that *milicianos* were filling most of the static garrison posts while the professional career officers did most of the fighting. This resentment was subsequently fuelled by the so-called Rebelo Decrees of July 1973 by which *milicianos* were granted accelerated promotion and seniority through having to complete only two-and-a-half and not the usual four semesters at the military academy. Ironically the measure had been designed to persuade the graduates to prolong their stay in the Army. The criticism was not stemmed by a further decree in August protecting the seniority of those above the rank of captain and the Armed Forces Movement (MFA) was founded in September 1973 as a direct result.

The fact that the eventual MFA programme was concerned with economic and social reforms within Portugal itself as well as professional grievances and that, similarly, Spinola's proclaimed counter-revolutionary programme in Guinea had also stressed the need for domestic reform, are indications that there were other domestic reasons for the coup of April 1974. There had always been left-wing opposition to Salazar and Caetano and to the war, the Armed Revolutionary Action (ARA) group, for example, having sabotaged helicopters at the Tancos air base near Lisbon in March 1971 and having been responsible for the sabotage of the coaster *Angoche* off Mozambique in April 1971. The large foreign-dominated commercial cartels such as CUF, which controlled the economic affairs of metropolitan Portugal as much as the colonies, were increasingly opposed not only to the war but to Salazarist economic doctrine. Significantly Spinola's celebrated critique of government policy, *Portugal and the*

Future, which led to his dismissal as Deputy Chief of Staff in March 1974 and helped to precipitate the coup, was published by a subsidiary of CUF. As industrialisation and modernisation proceeded in Portugal, it produced further internal strains such as increasing pressure on housing, while inflation rose from 16 per cent in 1971 to a staggering 72 per cent in February 1974 with the onset of the world oil crisis.

There had been considerable industrial unrest throughout 1973 and the continued emigration of Portuguese workers was in itself a judgement on what was in effect a colonial economy at home. Caetano had also proved far less liberal than had once been hoped. He had dropped talks which Spinola had been conducting with Senegalese leaders in mid-1972 which might have reached a settlement in Guinea, and remained absolutely committed to a role in Africa. Younger officers believed that it might now be best to abandon that role and older officers feared the Army being held responsible for any future disaster in the manner of the Goan episode of 13 years previously. In a sense the colonial campaigns had actually sustained the Salazarist system through military expenditure boosting the economy in the 1960s and by creating a semblance of national unity. They also ultimately contributed to its collapse, but one must bear in mind the role of resentment within the armed forces and other elements of Portuguese society at Portugal's economic backwardness, social inequities and political rigidity as well as the Army's professional grievances in bringing about the coup on 25 April 1974.

Some of the issues raised above can be further illustrated by concentrating in more detail on just one of the Portuguese campaigns — in this case, that in Mozambique. As already indicated, FRELIMO commenced its campaign on 25 September 1964 with its first foray into the Cabo Delgado region of northern Mozambique. Forewarned by the risings in Angola and Guinea, the Portuguese had already built up troop strength to approximately 16,000 men, although only five aircraft were initially available. By 1973–4 the Portuguese had deployed over 60,000 troops in the colony and the Air Force had grown to comprise 12 Fiat G-92's, 15 Harvard T6 converted trainers, 14 Alouette and two Puma helicopters, and five Nord-Atlas and seven DC-3 transport aircraft. Apart from this initial build-up of forces, the Portuguese also derived advantage from the natural barrier provided by the Rovuma River separating Tanzania from

Mozambique, as well as the problems of tribal demography placed in the way of FRELIMO's attempt to penetrate beyond the Mueda plateau heartland of the Makonde. Indeed in a population of over seven million Africans, the Makonde were but one of at least 19 tribes from nine major ethnic groups speaking 17 different languages. Although guerrilla activity was extended to the Niassa region in 1967, where the Nyanja tribe proved more co-operative towards FRELIMO, this was again offset by the willingness of the Moslem Yao to back the authorities. Guerrilla activity in the north consisted thereafter mostly of minelaying or hit-and-run attacks, known to the Portuguese as *flagelaçao* ('whipping burst').

The size of FRELIMO units steadily increased as the movement approached a maximum strength of perhaps 8000 guerrillas by 1967–8, but these numbers subsequently declined following the internal splits within FRELIMO of 1968–9 as represented by the disputes at the 2nd Party Congress and Mondlane's assassination. FRELIMO did succeed in closing many sisal plantations along the northern frontier, but had been mostly contained when Arriaga, who had become Portuguese ground force commander in May 1969, undertook a major offensive — Operation 'Gordian Knot' — in the dry season of 1970. Involving some 10,000 troops, the operation continued for seven months, in which time Arriaga claimed to have accounted for 651 guerrillas dead and a further 1840 captured, for the loss of only 132 Portuguese. He also claimed to have destroyed 61 guerrilla bases and 165 camps, while 40 tons of ammunition had been captured in the first two months alone. The co-ordination of heliborne assault after initial artillery and air bombardment, followed by mine clearance and consolidation on foot, undoubtedly severely damaged FRELIMO's infrastructure in the north. But, in the manner of such large-scale operations, it did not totally destroy the guerrilla capacity for infiltration, while Arriaga's critics maintained that his predecessors had achieved as much at less cost and effort. Further operations were thus required in the north such as Operations 'Garrotte' and 'Apio' during 1971.

Mention of Arriaga's operational statistics is, too, an indication of how difficult it is to reconcile conflicting claim and counter-claim in counter-insurgency. Between June 1970 and July 1971, for example, and overlapping 'Gordian Knot', FRELIMO claimed to have killed 1507 Portuguese troops, to have destroyed 261 vehicles, two aircraft and one helicopter and to have attacked 59 Portuguese posts, 17 bridges and four trains. FRELIMO figures rarely bore much relation to reality, while those of the Portuguese can at least be used to indicate the scale and nature of the fighting. In 1971, for example, the Portuguese undertook 3657 ground operations in the colony, 14,398 air missions, 28,060 air sorties and 675 naval missions and claimed 319 guerrillas taken and 1041 killed. In 1972 they undertook 6038 ground operations, 15,461 air missions and 627 naval missions and took 274 prisoners and captured 964 weapons.

Linked to Arriaga's drives in the northern regions was a more ambitious scheme completely to neutralise infiltration by constructing a 'human' barrier along the Rovuma line. Indeed Arriaga's definition of 'winning' the war was of being able immediately to detect and destroy any infiltrations into the colony. It was planned that the human element of Operation 'Frontier' would be provided both by new white settlers and resettlement of the native population, the settlements being linked by all-weather roads which would widen every 10 km (six miles) to form an airstrip. As in Angola, the idea of white *kibbutzim* or military colonists proved unsuccessful but 150 *aldeamentos* were constructed in the Cabo Delgado and Niassa regions with a population of some 250,000 people, one model *aldeamento* at Nangade housing 2500 alone. The programme was behind schedule in the north by 1974, rains having eroded many of the newly-laid road surfaces and the 2000 troops by then available being too few to prevent all infiltration. As already indicated, *aldeamentos* in Mozambique were generally unsuccessful in providing their inhabitants with the desired amenities — one in Niassa was actually sited in a swamp. But the eventual construction of 980 *aldeamentos* in the colony, with a population of around a million people, made nonsense of FRELIMO claims to be controlling over a million in the same areas.[13]

Although mostly associated with military rather than social or political solutions to insurgency, Arriaga's social-promotion programme was just as extensive as elsewhere in the colonies. About 50 per cent of the troops available were employed on it, building roads, medical centres, farms and cattle dips. The numbers of natives being educated in primary schools rose from 427,000 in 1964 to 603,000 by 1972 and the secondary school population from 19,761 to 44,368 in the same period. The Portuguese also developed the psychological aspects of counter-insurgency, over 1400 aerial broadcasts being made in 1971 and two-and-a-quarter million pamphlets dropped. In 1972 almost five million pamphlets were dropped. As already indicated, Arriaga claimed a particularly high rate for inducing captured insurgents to work against their former colleagues and the Portuguese generally exploited the divisions within FRELIMO, insinuating that the Makonde were cannon fodder for FRELIMO's 'southern' leadership. There were a number of important defections to the Portuguese, including Lazaro Kavandame in March 1969 and Dr Miguel Murrupa in 1970. The number of black or local troops in the Army was increased from approximately 40 per cent of the total to over 60 per cent between 1968 and 1974, including some 10,000 to 12,000 blacks serving in the elite GE, GEP or *Flechas*. Alone among Portuguese commanders, Arriaga also instituted political instruction for his own troops. Known as *mentalizacao* ('mentalisation'), the programme was conducted through unit officers, slogans and leaflets. In the case of black elite troops at least an hour each day was devoted to it. Arriaga, who was known to his troops as the 'Pink Panther', generally attempted to build up a personality cult, but his command was frequently criticised. He was accused of employing too many subordinates who had failed elsewhere and he was recalled in July 1973.

Part of the reason for Arriaga's recall was that, despite successes in the north, the Portuguese were facing increasing problems elsewhere. In particular FRELIMO had secured a major success by extending their campaign unexpectedly to the Tete region, the opening of their new 'front' being announced in March 1968. Far from both Army headquarters in Nampula and the seat of the civil administration at Lourenço Marques, Tete had been almost completely neglected by the Portuguese. It was believed to be inaccessible and under no real threat, despite the location there of the major Cabora Bassa project. Originating as a concept in 1957, construction work began in 1968 on a dam which would not only irrigate 1.5 million hectares (3.7 million acres) but provide the largest source of hydroelectric power in the entire continent. Infiltrating through Malawi and waging a selective terror campaign against tribal leaders, FRELIMO caught the Portuguese completely by surprise. A crash programme of *aldeamentos* was begun; by 1971 a military governor had been appointed and the dam ringed with 64 km (40 miles) of wire and minefields and a reported 15,000 troops. Unable to penetrate the new defences, FRELIMO resorted to attacking the 443 km (274 mile) route by which concrete was brought to the dam site. It is also possible that the change of tactics owed something to the benefit the dam might provide an independent Mozambique, although guerrilla spokesmen were careful to stress the value of the project to South Africa in terms of energy production. To leave the project entirely alone, of course, would also have created the impression that the Portuguese were invulnerable. For a short time the attacks succeeded in inducing the Portuguese to fly in concrete supplies, but work was never seriously interrupted and the project was six months ahead of schedule by the middle of 1972. So successful was its defence that a civilian governor was reappointed in September 1973 while the guerrillas were reduced to occasional and largely futile long-range bombardment by 122 mm rockets. Indeed the guerrillas generally were resorting to long-range weapons more and more throughout the colony, the 122 mm rocket first appearing in January 1973.

But the Portuguese appear to have concentrated to such an extent on defending Cabora Bassa that the remainder of the Tete region became a springboard from which ZANU (Zimbabwe African National Union) guerrillas, with whom FRELIMO closely co-operated, could attack into Rhodesia in December 1972. Much of the Rhodesian criticism of Portuguese efforts derived from their sense of the latter having let the guerrillas into Rhodesia by the 'back door', the Rhodesians preferring to co-operate with the DGS and *Flechas* rather than with an Army for which they had growing contempt. The other effect of the dam defences was to force guerrillas to move elsewhere and by 1972 FRELIMO was beginning to penetrate south and east from Tete, although in small numbers. The Vila Pery region was infiltrated during 1972 and the Beira region in the following year, with the Beira railway coming under attack as its importance to Zambia declined with the closure of the Rhodesian/Zambian frontier. Rail traffic ceased at night and armoured mine detectors or *zorras* were placed in front of trains. As the Rhodesian forces had increasing success in preventing sabotage of the Beira line, so FRELIMO switched their efforts to the Moatizo line into Malawi. The murder of the first white civilian sparked off riots against the Army in Beira in January 1974, culminating in an attack on the officers' mess and the forced closure of the Army's rest centre there. The demand by settlers, who had never enjoyed a close relationship with the Army, for more protection also enabled the DGS to exploit Army unpopularity in the prolonged struggle of Army and Police which had underlaid the entire campaign. By July 1974 guerrillas had penetrated the Zambezia region for the first time while, as already related, SAM missiles had made their appearance in the colony in March 1974. There had therefore been steadily escalating violence.

FRELIMO had always conducted terrorism against the native population — at least 689 deliberate assassinations had taken place between 1964 and February 1973, together with over 2000 woundings and 6500 abductions. In Tete alone 55 chiefs had been murdered during 1971. This terrorism was, moreover, increasing. Similarly Mozambique accounted for twice as many Portuguese casualties between November 1973 and January 1974 than either Angola or Guinea, and it is perhaps significant that most allegations and counter-allegations of atrocities came from Mozambique. Nevertheless it is important to stress that the Portuguese were still far from losing the campaign in the colony by 1974 and there is no reason to suppose that insurgency would have succeeded in the immediate future. As it was, the coup in Portugal brought about Portuguese withdrawal, transfer of power to FRELIMO being agreed in September 1974 for independence in June 1975.

The Portuguese may have had growing problems in Mozambique by 1974 but not serious enough to suggest military defeat. In Angola the war had become one of low-level stalemate, while the war with the most successful of the guerrilla groups, PAIGC in Guinea, had also been reduced to virtual stalemate. Throughout the three colonies the Portuguese retained control of all major routes and all urban areas, urban terrorism never having developed. As one analyst has remarked, the Army coup and its repercussions 'refuted the predictions of even the most astute forecasters'.[14] In military terms neither the Portuguese nor the guerrillas had won or lost, but, as in many other counter-insurgency campaigns, the ultimate outcome had been decided elsewhere.

Notes

1. P. Anderson, 'Portugal and the End of Ultra Colonialism', *New Left Review*, 16, 1962, p. 116; quoted in D. Porch, *The Portuguese Armed Forces and the Revolution* (Croom Helm, London and Stanford, 1977), pp. 13-14.

2. T. Gallagher, *Portugal: A Twentieth Century Interpretation* (Manchester Univ. Press, Manchester, 1983), p. 183.

3. A.J. Venter, *The Terror Fighters* (Parnell, Cape Town, 1969), p. 44; A.J. Venter, *Portugal's Guerrilla War* (Malherbe, Cape Town, 1973), p. 94.

4. The point is made by T.H. Henriksen in his articles 'Lessons from Portugal's Counter-Insurgency Operations in Africa', *Journal of the Royal United Services Institute*, 123/2, 1978, pp. 31–5 and 'Portugal in Africa: Comparative Notes on Counter-Insurgency', *Orbis*, Summer, 1977, pp. 395–412.

5. J.A. Marcum, *The Angolan Revolution*, vol. II (MIT, Cambridge, Mass., 1978), pp. 210–21.

6. P. Janke, *Southern Africa: End of Empires* (Conflict Studies, London, Paper no. 52, 1974), p. 5. See also G.J. Bender, 'The Limits of Counter-Insurgency: An African Case', *Comparative Politics*, 4/3, 1972, pp. 331–60 and B.F. Jundanian, 'Resettlement Programmes: Counter-Insurgency in Mozambique', *Comparative Politics*, 6/4, 1974, pp. 519–40.

7. Quoted in G.J. Bender, *Angola under the Portuguese* (Heinemann, London, 1978), p. 178.

8. D.L. Wheeler, 'African Elements in Portugal's Armies in Africa', *Armed Forces and Society*, 2/2, 1976, pp. 233–50.

9. B. Davidson, *The People's Cause* (London, 1981), p. 125.

10. Official figures from Cunha, Arriaga, Rodriques and Marques, *A Vitoria Traida* (Lisbon, 1977), p. 68; quoted in Davidson, *People's Cause*, p. 179 and B. Davidson, *No Fist is Big Enough to Hide the Sky* (London, 1981), p. 160. Venter's calculation is in *Portugal's Guerrilla War*, p. 75 and A.J. Venter (ed.), *Africa at War* (Old Greenwich, Conn., 1974), p. 75.

11. Quoted in B. Davidson, J. Slovo and A.P. Wilkinson (ed.), *Southern Africa* (Penguin Books Ltd., Harmondsworth, 1976), p. 19.

12. D.L. Wheeler, 'The Portuguese Army in Angola', *Journal of Modern African Studies*, 7/3, 1969, pp. 425–39.

13. A.R. Wilkinson, 'Angola and Mozambique: The Implications of Local Power', *Survival*, 16/5, 1974, pp. 217-27.

14. T.H. Henriksen, 'Some Notes on the National Liberation Wars in Angola, Mozambique and Guinea-Bissau', *Military Affairs*, 41/1, 1977, pp. 30-6.

References

A. *General Works*
Abshire, D.M. and Samuels, M.A. (ed.), *Portuguese Africa: A Handbook* (Praeger, London, 1969)
Gallagher, T. *Portugal: A Twentieth Century Interpretation* (Manchester Univ. Press, Manchester, 1983)
Newitt, M. *Portugal in Africa: The Last Hundred Years* (Hurst, London, 1981)

B. *Angola and Mozambique — General Accounts*
Henriksen, T.H. *Mozambique: A History* (Rex Collings, London and Cape Town, 1978)

Munslow, B. *Mozambique: The Revolution and Its Origins* (Longman, London, 1983)

Wheeler, D. and Pelissier, R. *Angola* (Praeger, London, 1971)

C. *Accounts from the Perspective of the Guerrillas*

Barnett, D. *With the Guerrillas in Angola* (Liberation Support Movement, Seattle, 1970)

Chaliand, G. *Armed Struggle in Africa* (Monthly Review Press, London and New York, 1969)

Cornwall, B. *The Bush Rebels* (Andre Deutsch, London, 1973)

Davidson, B. *The Liberation of Guinea* (Penguin Books Ltd., Harmondsworth, 1969). Since reprinted in a second edition as *No Fist is Big Enough to Hide the Sky* (Zed, London, 1981)

Davidson, B. *Walking 300 Miles with Guerrillas through the Bush of Eastern Angola* (Pasadena, Munger Africana Notes no. 6, 1971)

Davidson, B. *In the Eye of the Storm* (Longman, London, 1972)

D. *Accounts or Publications by Guerrilla Leaders*

Cabral, A. *Revolution in Guinea* (Monthly Review Press, New York, 1969)

McCulloch, J. *In the Twilight of Revolution* (Routledge and Kegan Paul, London, 1983)

Mondlane, E. *The Struggle for Mozambique* (Penguin Books Ltd., Harmondsworth, 1969)

Wolfers, M. (ed.) *Unity and Struggle* (Heinemann, London, 1980)

E. *Accounts of the Campaigns, Broadly Sympathetic to the Guerrillas*

Barnett, D. and Harvey, R. (ed.) *The Revolution in Angola: MPLA Life Histories and Documents* (Bobbs, New York, 1972)

Davidson, B. *The People's Cause* (Longman, London, 1981)

Davidson, B., Slovo, J. and Wilkinson, A.R. (ed.), *Southern Africa* (Penguin Books Ltd., Harmondsworth, 1976)

Ehrmark, A. and Wastberg, P. *Angola and Mozambique: The Case against Portugal* (Pall Mall Press, London, 1963)

Gibson, R. *African Liberation Movements* (Oxford Univ. Press, Oxford, 1972)

Humbaraci, A. and Muchnik, N. *Portugal's African Wars* (Macmillan, London, 1974)

F. *Accounts of Observers with the Portuguese Forces*

Biggs-Davidson, J. 'Portuguese Guinea: A Lesser Vietnam', *NATO's Fifteen Nations*, 13/5, Oct.–Nov. 1968, pp. 21–5

Bruce, N. *Portugal's African Wars* (Conflict Studies, London, Paper no. 34, 1973). Later expanded into *Portugal: The Last Empire* (David and Charles, Newton Abbott, 1975)

Calvert, M. 'Counter-Insurgency in Mozambique', *Journal of the Royal United Services Institute*, 118/1, 1973, pp. 81–4

Morris, M. *Armed Conflict in Southern Africa* (Howard Timmins, Cape Town, 1974)

Swift, K. *Mozambique and the Future* (R. Hale, London, 1975)

Venter, A.J. *The Terror Fighters* (Parnell, Cape Town, 1971)

Venter, A.J. *Portugal's War in Guinea-Bissau* (Pasadena, Munger Africana notes No. 19, 1973)

Venter, A.J. *Portugal's Guerrilla War* (Malherbe, Cape Town, 1973)

Venter, A.J. *Africa at War* (Devin-Adair, Old Greenwich, Conn., 1974)

Venter, A.J. *The Zambezi Salient: Conflict in Southern Africa* (R. Hale, London, 1975)

162   *The Portuguese Army*

G. *Accounts by Portuguese Commanders*
de Arriaga, K. *The Portuguese Answer* (Tom Stacey, London, 1973)
de Spinola, A. *Portugal and the Future* (Johannesburg, 1974)

H. *Accounts of Portuguese COIN*
Alberts, D.J. 'Armed Struggle in Angola' in B. O'Neill *et al.* (ed.), *Insurgency in the Modern World* (Westview Press, Boulder, Colorado, 1980), pp. 235–68
Bender, G.J. 'The Limits of Counter-Insurgency: An African Case', *Comparative Politics*, 4/3, 1972, pp. 331–60
Bender, G.J. *Angola under the Portuguese* (Heinemann, London, 1978)
Henriksen, T.H. 'Some Notes on the National Liberation Wars in Angola, Mozambique and Guinea-Bissau', *Military Affairs*, 41/1, 1977, pp. 30–6
Henriksen, T.H. 'People's Wars in Angola, Mozambique and Guinea-Bissau', *Journal of Modern African Studies*, 14/3, 1976, pp. 377–99
Henriksen, T.H. 'Portugal in Africa: Comparative Notes on Counter-Insurgency', *Orbis*, Summer 1977, pp. 395–412
Henriksen, T.H. 'Lessons from Portugal's Counter-Insurgency Operations in Africa', *Journal of the Royal United Services Institute*, 123/2, 1978, pp. 31–5
Jundanian, B.F. 'Resettlement Programmes: Counter-Insurgency in Mozambique', *Comparative Politics*, 6/4, 1974, pp. 519–40
Marcum, J.A. *The Angolan Revolution* (MIT, Cambridge, Mass., 1969 and 1978; 2 vols)
Middlemas, K. *Cabora Bassa* (Weidenfeld and Nicholson, London, 1975)
Wheeler, D.L. 'The Portuguese Army in Angola', *Journal of Modern African Studies*, 7/3, 1969, pp. 425–39
Wheeler, D.L. 'African Elements in Portugal's Armies in Africa', *Armed Forces and Society*, 2/2, 1976, pp. 235–50
Wilkinson, A.R. 'Angola and Mozambique: The Implications of Local Power', *Survival*, 16/5, 1974, pp. 217–27

I. *The 1974 Coup and its Aftermath*
Fields, R.M. *The Portuguese Revolution and the Armed Forces Movement* (Praeger, New York, 1975)
Porch, D. *The Portuguese Armed Forces and the Revolution* (Croom Helm, London and Stanford, 1977)
Sobel, C.A. *Portuguese Revolution, 1974–76* (Facts on File, New York, 1976)
Sunday Times Insight Team, *Insight on Portugal: The Year of the Captains* (Andre Deutsch, London, 1975)

6   THE RHODESIAN ARMY:
COUNTER-INSURGENCY, 1972–1979

Ian F.W. Beckett

Modern counter-insurgency is rarely a purely military problem for a government and its Security Forces. Of this basic truism, the experience in Rhodesia between 1966 and 1979 affords a significant example. Not only were the efforts of the Rhodesian Security Forces frequently directed towards particular political goals, but their ultimate failure to contain insurgency at an acceptable level derived to a large extent from external political pressures over which they had little control.

In a real sense, Rhodesia was the creation of private enterprise rather than the British government, the British South Africa Company of Cecil Rhodes annexing Mashonaland in 1890 and Matabeleland in 1893. The Company continued to run the administration until, following a referendum of the white settlers which indicated their long-standing disillusionment with such control, Southern Rhodesia as it was then known became a self-governing colony in 1923. Far more prosperous than either Northern Rhodesia or Nyasaland, Southern Rhodesia effectively dominated the Central African Federation into which it entered with these neighbours in 1953. The Federation collapsed in 1963, with Northern Rhodesia and Nyasaland becoming the independent black states of Zambia and Malawi respectively in the following year. The larger white settler population in Southern Rhodesia rejected the concept of majority rule, a determination reinforced by the spectre of chaos in the Belgian Congo in 1960 and by the urban unrest in Southern Rhodesia itself which followed the rejection by the growing black nationalist movement of the proposed 1961 constitution, despite its greater participatory role for the African.

The nationalist movement had developed in the 1950s with Joshua Nkomo's African National Congress being established in 1957. Banned on a number of occasions, Nkomo's group re-emerged under different titles, becoming the National Democratic Party in 1960 and the Zimbabwe African People's Union

(ZAPU) in 1962. The nationalists were moving towards the advocacy of violence to achieve their political aims at the very time when white resistance was symbolised by the sweeping electoral victories of Ian Smith's Rhodesian Front Party in December 1962. The advent of a Labour administration in Britain in 1964, dedicated to majority rule, catapulted all sides closer to confrontation and, on 11 November 1965, Ian Smith issued a Unilateral Declaration of Independence (UDI).

The war that evolved in Rhodesia therefter has to be seen in the context of continuing political and diplomatic activity aimed at securing Rhodesian acceptance of majority rule and the end of rebellion against the Crown. The British rejected the use of force, although they did resort to largely ineffectual economic sanctions, including the so-called Beira Patrol off the coast of Portuguese Mozambique from 1965 to 1974, when the latter became independent. Similarly, British troops were stationed in Bechuanaland (later Botswana) from 1965 to 1967 to guard a BBC transmitter at Francistown from possible Rhodesian sabotage. In December 1966 Smith met the British prime minister, Harold Wilson, for talks aboard HMS *Tiger* and there were more negotiations aboard HMS *Fearless* in September 1968. The subsequent Conservative government sent the abortive Pearce Commission to Rhodesia from January to May 1972 to test the acceptability of new Anglo-Rhodesian proposals on a constitution.

Following the collapse of Portuguese control in Mozambique, Rhodesia not only became more exposed to guerrilla infiltration, but also suffered increasing pressure from the South African prime minister, John Vorster, to reach an accommodation with the guerrillas. A brief ceasefire came into effect in December 1974 and, although this failed, Vorster and Zambia's president, Kenneth Kaunda, arranged negotiations between Ian Smith and nationalist leaders on the Victoria Falls bridge in August 1975. There were further talks between Smith and Nkomo in early 1976 and in September of that year, under considerable South African pressure, Smith conceded the principle of majority rule within the context of an overall agreement worked out by Vorster and the United States Secretary of State, Henry Kissinger. A conference at Geneva from October 1976 to January 1977, however, failed to produce a settlement acceptable to all parties, and further proposals put forward by the British and US

Map 6.1: Rhodesia 1972–9

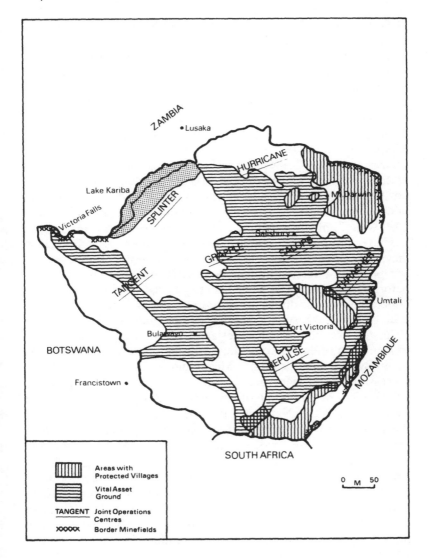

governments in September 1977 also came to nothing. Ian Smith then reached an internal settlement with three nationalist leaders — Bishop Abel Muzorewa, the Reverend Ndabaningi Sithole and Chief Jeremiah Chirau — in March 1978, by which Muzorewa and Sithole entered a transitional government. This did not, however, rule out further negotiations between Smith and Nkomo in Lusaka in August 1978. In April 1979, as a result of internal elections, Muzorewa became the first black prime minister of Rhodesia. The republic declared in March 1970 was formally brought to an end in June 1979 with the creation of Zimbabwe-Rhodesia. The final political turn of events was the Lancaster House Conference in London between September and December 1979, which resulted in a British-supervised ceasefire on 28 December 1979 and a transitional British administration under Lord Soames as governor. Elections were held in February 1980 with Zimbabwe gaining full legal independence in April.

These complicated political events between 1965 and 1980 inevitably affected the conduct of the war inside Rhodesia and across its frontiers, although large-scale conflict did not occur before December 1972. Thus Security Force operations could be undertaken to put direct pressure upon the guerrillas in order to achieve political results in the wider diplomatic field. In October 1976, for example, the Rhodesians frustrated guerrilla attempts to launch an offensive coinciding with the Geneva Conference by themselves striking deep into Mozambique. Similarly, the highly-successful Rhodesian attack on New Chimoio (Operation 'Miracle') in Mozambique in September 1979 put pressure on the Zimbabwe African National Liberation Army (ZANLA) during the Lancaster House Conference. Moreover, there was a whole series of attacks on economic targets in both Mozambique and Zambia, designed to compel the guerrillas' hosts — Kaunda and Samora Machel, President of Mozambique — to ensure that their clients adopted a more positive approach to the negotiations. In September 1979, for example, Rhodesia suspended Zambian maize shipments on Rhodesian railways, Zambia having been forced by economic pressure to reopen its frontiers with Rhodesia in 1978. In October 1979 Zambia's own railway system came under Rhodesian attack, while it has been estimated that Mozambique suffered over 26 million dollars' worth of damage in 1979.[1] In February 1979 Angolan targets had been bombed by the Rhodesian Air Force to frustrate any guerrilla build-up prior to the internal elections.

The fact that the ZANLA guerrillas and those of the Zimbabwe People's Revolutionary Army (ZIPRA) operated from sanctuaries in other countries, resulted in further political complications. In January 1973 Rhodesia closed its frontier with Zambia, with the exception of copper shipments, as a direct result of the escalation of guerrilla activity in the north-east of Rhodesia. The Rhodesians subsequently reopened the frontier but, as indicated above, Zambia then declined to do so until 1978. Similarly, Mozambique closed its frontiers with Rhodesia in March 1976 and by the end of the war only the 222 km (138 mile) frontier with South Africa out of a total frontier length of 2964 km (1841 miles) was entirely free of infiltration. The Rhodesians had, in fact, begun operating up to 100 km (62 miles) inside Mozambique in co-operation with the Portuguese as early as 1969. The first large-scale cross-border raid was not launched, however, until August 1979 (Operation 'Eland'). Such raids occurred frequently thereafter, often coinciding with the imminent approach of the rainy season in November and hitting the guerrilla concentrations that would have attempted to infiltrate under favourable climatic conditions that restricted Rhodesia's monopoly of air power. Physical difficulties as well as political restraint precluded large-scale raids into Zambia until October 1978, when the first took place in direct response to the shooting down of a Rhodesian Viscount civil airliner a month previously by ZIPRA, who had then massacred the survivors. A second airliner was shot down in February 1979, eliciting the air strike into Angola although, as already indicated above, the operation fulfilled other requirements as well. There was also an attempt to kill Nkomo in Lusaka in April 1979 while, earlier in the same year, Rhodesian forces had sunk the Kasangula ferry which was Botswana's only link with Zambia. But, just as the neighbouring black states were to some extent dependent upon Rhodesia's railways for their survival, Rhodesia itself after UDI was equally dependent upon external sources.

With the withdrawal of the Portuguese, Rhodesia's lifeline lay through South Africa, but it is clear that Vorster sacrificed Rhodesian whites in the cause of wider detente with black states. Thus although elements of the South African police were committed to assist the Rhodesians in 1967, they were withdrawn by August 1975 to facilitate the attempt by Kaunda to get the nationalists to negotiate and, equally, to put pressure on Smith to do the same. In fact, a number of South African pilots and technicians remained in Rhodesia, but they were also recalled following the first major raid into Mozambique in August 1976 which Vorster feared would jeopardise relations with Machel. Furthermore, the South African foreign minister then broadcast his government's support for majority rule in Rhodesia which, as two recent historians of the war have written, 'pulled the rug' from under Ian Smith.[2] Subsequently, Vorster's successor as prime minister, P.W. Botha, went some way towards reversing the situation by lending the Rhodesians military equipment and personnel and committing South African troops to defend key points such as the Beit Bridge which linked Rhodesia and South Africa across the River Limpopo.

Vorster had feared the consequences of any escalation in the war between Rhodesia and its neighbours and there were inevitably clashes between Rhodesian forces and those of the black states. On one notable occasion in September 1979 during the attack on New Chimoio, Rhodesian Eland armoured cars, a version of the Panhard, engaged Soviet-supplied T-34 tanks of the Mozambique Army (FPLM). On such raids the Rhodesians invariably had two Hawker Hunter jets armed with 68 mm rockets avilable as an anti-tank reaction force.[3] Curiously, the Rhodesians themselves also had some Soviet T-55 tanks, which had been landed in South Africa instead of the intended destination of Uganda when Idi Amin's regime fell in 1979. But incursions into neighbouring states also carried the possibility of clashing with the wide variety of foreign nationals — Chinese, Russians, Cubans and so on — who advised the guerrillas. In the case of the struggle for New Chimoio, for example, East German advisers fought with ZANLA guerrillas. The guerrillas were, of course, also sustained by many other external organisations, including the Organisation of African Unity (OAU), the World Council of Churches and the Third World lobby in the United Nations.

The involvement of both Chinese and Soviet advisers with the guerrillas is in itself an indication that the struggle inside Rhodesia was yet further complicated by the existence of deep rivalries within the nationalist movement. As early as 1963 Sithole had split away from Nkomo's ZAPU to form the Zimbabwe African National Union (ZANU). To a large extent the split was along tribal lines, ZAPU being based on the minority Ndebele of western Rhodesia and ZANU on the majority Shona of eastern Rhodesia. This contributed to the tendency of the two organisations to operate in what might be termed the area of their 'natural' support. Thus ZAPU and its military wing, ZIPRA, operated out of Zambia and Botswana, while ZANU and its military wing, ZANLA, operated out of Mozambique. While the early guerrillas of both organisations had trained in diverse or even the same countries overseas, ZIPRA came increasingly to reflect Soviet orthodoxy and ZANLA to reflect Chinese theories of rural guerrilla warfare. Thus a feature of the war after 1972 was the reluctance of Nkomo to commit large numbers of his ZIPRA forces to Rhodesia, preferring to retain them in Zambia for a Soviet-style conventional assault at an appropriate moment. Indeed, a number of Rhodesian spoiling operations in late 1979 were specifically mounted to disrupt the ZIPRA build-up, including the destruction of key road bridges which might have been used to throw ZIPRA armour across the Zambezi.

ZANLA also briefly considered a conventional assault in 1979 to establish a provisional government inside Rhodesia, but for the most part the approach of the two groups was markedly different. This made co-operation between ZANLA and ZIPRA difficult and there were further break-aways such as that of James Chikerema, who left ZAPU to form the Front for the Liberation of Zimbabwe (FROLIZI), forcing a hasty and temporary junction of ZIPRA and ZANLA in a Joint Military Command in 1972. Bishop Abel Muzorewa, who had emerged during the Pearce Commission as a nationalist of some authority, and his United African National Council (UANC) was recognised by the OAU in 1974 as a means of uniting the disparate guerrilla struggle. Muzorewa joined together with Nkomo, Sithole and Chikerema to form a Zimbabwe Liberation Council in 1975, while ZIPRA and ZANLA were forced by their black African hosts to create a unified army in the shape of the Zimbabwe People's Army (ZIPA). This unity quickly faded, the subsequent union of ZAPU and ZANU in the so-called Patriotic Front for the purpose of attending the Geneva Conference in 1976 never resulting in any actual military unity between ZIPRA and ZANLA. Within ZANU, Sithole was by now being out-manoeuvred by more radical elements and, after his release from detention inside Rhodesia in 1974, Robert Mugabe became the dominant figure. Thus Sithole, still heading a group he called ZANU, came together with Muzorewa's UANC and Chief Chirau's insignificant Zimbabwe United People's Organisation in accepting the internal settlement in 1978. Their three parties contested the internal elections in April 1979, all seeking the Shona vote rather than that of the Ndebele. Subsequently, Chikerema deserted the UANC to form yet another faction — the Zimbabwe Democratic Party. Throughout the war, therefore, there were divisions among the nationalists that could be successfully exploited by the Rhodesians as the internal settlement indicated only too clearly. On occasions there were clashes between rival guerrillas and a number of major internal upheavals such as the 'Nhari Rebellion' in ZANLA in December 1973 and the assassination of one of ZANLA's leaders, Herbert Chitepo, in Lusaka in March 1974 which led to ZANLA's virtual expulsion from Zambia.

At the time when insurgency first began, the complicated nature of the divisions among the nationalists was not apparent and the nature of the insurgency itself was limited. The first white man was killed by a so-called ZANU 'Crocodile Commando' in July 1964, but the first systematic attempt to infiltrate guerrillas into Rhodesia did not occur until April 1966, when a group of 14 ZANU guerrillas crossed into the country from Zambia. Over the course of the next two years a variety of guerrilla columns from ZANU, ZAPU and, on occasions, ZAPU guerrillas co-operating with the South African branch of the African National Congress, were comfortably contained and successfully eliminated by the Rhodesian Security Forces to such an extent that virtually all insurgency ceased for the next four years. The ease with which the guerrillas had been defeated did, however, have subsequent repercussions since it was largely seen as a police action and was controlled by Rhodesia's British South Africa Police (BSAP). The Rhodesian Army was rarely used, even though the BSAP was frequently operating as a conventional military force with patrols, sweeps and supported by helicopters.[4] Similarly, the BSAP Special Branch was especially prominent, its network of informers working well since the local African population of the Zambezi valley had little sympathy for the guerrillas. In any case the valley was an inhospitable environment and few guerrillas penetrated beyond it. Where military support had been required, temporary brigade areas were established with a Joint Operations Centre (JOC) involving military and police representatives as well as civil commissioners from the Department of Internal Affairs.

When insurgency developed once more, with the opening of ZANLA's new front in the Centenary district of the north-east in December 1972, there was a natural tendency to persist with previous practices. Beyond the local JOCs, the chain of command therefore stretched upwards through provincial JOCs, a Joint Planning Staff (JPS), and a Deputy Minister in Ian Smith's office (from 1974), to the Security Council of the Rhodesian Cabinet. In September 1976 a War Council replaced the Security Council and in March 1977 a Combined Operations Headquarters (Comops) replaced the JPS. In theory the creation of Comops should have enabled the Security Forces to develop a well co-ordinated strategy for the prosecution of the war. In reality, the command and control system failed at a number of levels. For one thing, there was increasing friction between Army and Police as the escalation of the war led to the replacement of BSAP personnel by the military in positions of responsibility on JOCs. In 1973 the JOC in the northeast was converted into a permanent operational brigade area — 'Hurricane'. This was followed by the establishment of 'Thrasher' and 'Repulse' in 1976, 'Tangent', 'Grapple' and 'Splinter' in 1977, and 'Salops' in 1978. With the exception of the latter, which remained a largely administrative creation under BSAP control, the other JOCs were now chaired almost as a matter of course by the Army.

The Army was also increasingly critical of other civilian government agenices, notably Internal Affairs which it held responsible for failing to perceive the nature of the growing ZANLA threat in the northeast prior to its eruption in 1972. Comops offered the possibility of reconciling differences but it never had effective control over civil affairs and ministries like Internal Affairs and Law and Order which had a considerable contribution to make to the war effort. Moreover, Comops became entangled in the day-to-day conduct of the war rather than in planning long-term strategy. Its commander, Lieutenant-General Peter Walls, also assumed command of all offensive and special forces as well as responsibility for all external operations. This left the Army commander, Lieutenant-General John Hickman, commanding only black troops and white territorials, while his staff were deprived of any real function at all. One commentator with first-hand knowledge of the system has claimed that the *de facto* commander of the Army in these circumstances was the Brigadier of Comops.[5] Walls sought further clarification of his powers but was to be disappointed, although it has been claimed that by 1979 he was the most powerful man in Rhodesia[6] and certainly the command structure as a whole was streamlined after the internal settlement, to exclude Muzorewa and Sithole from effective influence. At the same time Smith's influence also waned and he was on bad terms with Walls.

With division at the top of the system, it is not unlikely that this will be magnified at lower levels and such was the case in Rhodesia. There was, for example, an attempt to co-ordinate the Rhodesian Special Air Service (SAS) and the Selous Scouts with the establishment of a Special Forces Headquarters in July 1978. However, this fell foul of inter-unit rivalry and was eventually confined to administering the black Security Force Auxiliaries (SFAs) that came into existence after the internal settlement. The rivalry between the Army and the BSAP was also apparent in the attempted co-ordination of intelligence. Prior to 1972 intelligence was firmly a BSAP responsibility and of its Special Branch in particular, as was so frequently the case in British or former British territories. The Army had no intelligence network of its own, but the lack of real insurgency simply did not necessitate it. This was to change with the escalation of conflict in December 1972. In the northeast, which had been generally neglected by the Rhodesian administration at all levels, Special Branch's traditional reliance upon a handful of picked informers proved hopelessly inadequate. When the Army subsequently formed its own Military Intelligence Department in 1973, however, Special Branch regarded it with suspicion. The Department had no effective access to captured insurgents until 1978 and was generally confined to gathering external intelligence, largely through its radio interception service. There was also no Intelligence Corps formed within the Army until July 1975. Similarly, Special Branch initially controlled the special intelligence-gathering units which were raised by Major Ron Reid-Daly between November 1973 and January 1974. Subsequently named the Selous Scouts in March 1974, the regiment came under Comops control in 1977. There is some evidence of friction between the Selous Scouts and the Army, the attempt by Reid-Daly to recruit black servicemen from the Rhodesian African Rifles (RAR) being persistently resisted. Equally, the blowing of the cover of the Selous Scouts' first 'pseudo' operation in January 1974 by a Special Branch officer led to friction between the BSAP and the Scouts. A further indication of some of the tensions within the armed forces was the allegation in 1979 that the Selous Scouts were more intent on ivory poaching than killing guerrillas in areas frozen to operations by other members of the Security Forces. The Army's Intelligence Department bugged Reid-Daly's telephone and, amid the reverberations, Hickman was sacked as Army commander and Reid-Daly court-martialled. Reid-Daly was reprimanded and retired.

The lack of co-ordination of both command and intelligence was an important drawback to the Security Forces since they would always be stretched numerically, given the view of the Rhodesian authorities that the effective ceiling on manpower was the available white male population. Prior to the war, Rhodesia's regular forces were small and in 1968 still amounted to only 4600 men in the armed forces and 6400 in the BSAP, excluding reserves in both cases. By 1978 the whites numbered only 260,000 in a total population of some 6.9 million and, as the war progressed, white emigration — the 'chicken run' — outpaced immigration. Between 1960 and 1979 some 180,000 whites entered the country but 202,000 left, a net loss of over 13,000 whites in 1978 being the highest recorded. Under the 1957 Defence Act, young white males were liable to a six-week period of training in the Rhodesia Regiment, a territorial formation, followed by a reserve commitment. By 1966 the basic term of national service had increased to 245 days. In December 1972 national service for all whites as well as Asians and coloureds (who numbered about 30,000) between the ages of 18 and 25 was increased to a full 12 months, while the period to be spent in the reserve was increased from four to six years. In February 1974 the size of the annual intake was doubled and in November restrictions placed on the ability of those liable to military service to leave the country. In May 1976 the period of liability for territorials was increased indefinitely and the initial term of national service increased from 12 to 18 months. In January and February 1977 the net was widened still further with those aged between 24 and 38 compelled to do 190 days' service per annum,

174 The Rhodesian Army

those aged between 38 and 50 made liable to 70 days' service a year, and those aged over 50 encouraged to volunteer for the BSAP reserve, which required 42 days' service per annum for this age bracket. In September student deferments were cancelled and rewards advertised for those willing to extend their terms of service. It should be noted that the term of service of the older age groups was not continuous but completed as a number of tours through the year, such as six weeks on and six weeks off to try and minimise economic disruption. In January 1978 the deferment of two years for new immigrants of military age was reduced to just six months, although in October the term of service for those aged between 18 and 25 was once more reduced to 12 months. The ultimate measure of white conscription was introduced in January 1979 when those aged between 50 and 59 were made liable to six-weeks' service per annum, the new entrants being referred to as 'Mashford's Militia' after a well-known Salisbury funeral parlour.

Despite the increasing demands made upon white manpower, the great majority of the personnel of the Security Forces remained black. Until 1979 they were also all volunteers and there was no shortage of recruits, particularly among the Karanga tribe. Accordingly, the RAR added a second battalion in 1974, a third in 1977 and a fourth in 1978, the establishment of the latter raising the proportion of black servicemen from some 66 per cent of the whole to around 70 per cent. Approximately 75 per cent of the BSAP were also black, including most of the Police Support Units (PSU), popularly known as 'Black Boots'. Africans were attracted not only by good pay, housing, educational facilities and health care but also by traditional bonds of family service to the state. By 1979, too, there were 30 black commissioned officers in the Army. There is little evidence of disciplinary problems among black service personnel, although it would appear that some opposed the Anglo-Rhodesian proposals tested by the Pearce Commission and that the majority of the RAR probably voted solidly for Mugabe in the 1980 elections.

There was therefore no pressure for African conscription until after Muzorewa and Sithole joined the transitional government in 1978. In October it was announced that conscription would be introduced for educated Africans between the ages of 18 and 25 in January 1979. The measures were then extended to all educated Africans between 16 and 60 in August 1979, but there is evidence of some opposition to conscription among Africans and the scheme had not been fully implemented by the time the war ended. Somewhere between 1000 and 2000 foreigners also served with the Rhodesian Security Forces during the war, while the South African presence between 1967 and 1975 amounted to perhaps 2000 to 3000 men at most. In theory the Security Forces thus had large numbers of men available by the end of the war, but the requirements of the economy meant that only a relatively small proportion could be deployed at any one time. This usually amounted to about 25,000 men, although in the run-up to the internal elections in April 1979, some 60,000 men were deployed in the field, but only for a short period. It was only the establishment of the SFAs after the internal settlement that enabled the Rhodesians to reach even this total.

The lack of manpower tended to imply that there was little administrative 'tail' to the Security Forces, since traditionally most support functions had been undertaken by African labourers. The majority of the white national servicemen, especially older age groups, were also placed in a variety of more or less static roles such as holding units, police reserve units, the Guard Force created in February 1976 to assist the defence of protected villages (PVs), and the Defence Regiment formed in 1978 to guard important installations and communications. The principal strike formations were the regulars of the all-white Rhodesian Light Infantry (RLI), the white SAS, and the mixed-race Selous Scouts. While the RLI and the RAR provided the men for the 'Fire Forces' inside Rhodesia, the SAS and Scouts were available for external operations. There were also some other specialist units for counter-insurgency. The BSAP, for example, had Police Anti-Terrorist Units (PATU) as well as the PSUs, specialist anti-stock theft teams and SWAT (Special Weapons and Tactics Teams) which were designed to contain urban terrorism. The latter, however, was relatively limited, the most successful urban guerrilla operations being the bomb planted in the Salisbury branch of Woolworths in August 1977 and the rocket attack on the capital's oil storage depot in December 1978. The Ministry of Internal Affairs also fielded African District Security Assistants (DSAs) from 1976 for security duties in PVs. Another specialist Army unit was the Grey's Scouts, a mixed-race mounted unit often used to patrol border minefields.

The manpower shortage also had repercussions in terms of strategy in frontier areas and to prevent guerrilla infiltration into the interior. Ironically, the white urban areas and farms were surrounded by the African Tribal Trust Lands (TTLs) in a manner approximating to the Maoist guerrilla theory that the countryside dominated by insurgents should surround the cities. The need to prevent infiltration was an additional reason for striking at guerrilla concentrations outside Rhodesia. The Rhodesian forces were, in fact, well suited to counter-insurgency and had begun a systematic study of the subject in the 1950s. Some 50 per cent of all regular training was in the form of small-unit operations. There was also a reservoir of expertise from direct experience of British counter-insurgency operations. The Rhodesian Far East Volunteer Unit had served in Malaya during the Emergency in the 1950s; the then single battalion of the RAR had served in Malaya from 1956 to 1958; and Rhodesia's SAS had begun life as 'C' (Rhodesia) Squadron of the Malayan Scouts, later named 'C' (Rhodesia) Squadron of the British SAS, and had served both in Malaya and in Aden. The Rhodesian Air Force had also sent elements to Kuwait and Aden between 1958 and 1961. Indeed, it was sometimes alleged that there was a 'Malayan' clique within the armed forces, Walls having commanded the Rhodesian SAS squadron in Malaya. More recent experience was also available, the Selous Scouts being modelled to some extent on Portugal's *Flechas* whom Reid-Daly had studied. There was also close study of Israeli techniques, particularly in terms of external operations.[7]

Yet, despite the expertise available, the crucial lack of co-ordination in command and control prevented the development of the kind of distinct overall strategy that had characterised the British operations with which the Rhodesians were so familiar. Comops appeared after its creation in 1977 to abandon the generally-defensive reaction to guerrilla infiltration of earlier years in favour of a strategy of mobile counter-offensive. But, in the absence of sufficient numbers of men on the ground, the success of the counter-offensive largely depended upon inflicting high kill ratios. No real attempt could be made to hold cleared areas until the SFAs became available and it was not until 1979 that an area defence system was adopted, based on firmly holding 'Vital Asset Ground' corresponding to the white areas of Rhodesia.[8] This did not mean that some areas were tacitly abandoned to the guerrillas since elite groups such as the Selous Scouts would make periodic forays and the remaining ground of 'tactical importance' outside the vital asset ground, primarily the TTLs and game parks, became available for locating and destroying guerrillas at will. It was, however, late in the day before such a co-ordinated strategy was evolved and it has been suggested that the apolitical nature of the Rhodesian armed forces prevented them from seriously coming to terms with the political aspects of guerrilla insurgency.[9] There was never any real attempt at political indoctrination or instruction within the Rhodesian armed forces and to the end of the war guerrilla insurgency tended to be regarded as a military rather than a political problem to which military solutions alone should be applied.

Tactical considerations also tended to be affected by manpower restraints. Large numbers of men were required in static positions guarding installations, the vitally important railways, PVs and white farms. A reflection of this fact was the development of the 'Fire Force' concept which sought to offset lack of men through the concentration of firepower and mobility. If guerrillas were located by ground patrol or other means, a Cessna Lynx carrying fragmentation and concussion bombs or napalm would attack them. Four helicopters would then be deployed, each carrying a 'stick' of four or five men to drive the guerrillas back on 15 or 16 paratroopers dropped at low level from a Dakota C-47 transport. Four Fire Forces were available, two manned by the RLI and two by the RAR, with the men regularly rotated. Much depended upon flying time from base and, increasingly, on the number of requests for assistance. By 1978 a delay of several hours was common when earlier reaction had been almost instantaneous, since each Fire Force was being used two or three times a day in varying parts of the country. By mid-1979 the Fire Forces were said to be accounting for three quarters of all guerrilla casualties inside Rhodesia, but the Selous Scouts equally claimed that they were responsible for 68 per cent of all guerrilla kills, their early role in pseudo operations having been supplanted steadily by their deployment in a hunter-killer role. There appears to have been some resentment on the part of those who did the tracking on the ground only to see the reward of their labour claimed by other hands in the shape of the Fire Forces, but generally bifurcation was not a significant problem in the armed forces. The size of the Fire Force sticks was determined by the capacity of the Alouette helicopters with which the Rhodesians were primarily equipped, some 66 being available by 1979. It was also the size of sticks deployed in ground operations of a more conventional kind, companies being divided in this way to cover more ground while keeping in touch through the liberal distribution of personal radios. Subsequently, 11 or 12 Bell Huey helicopters with a greater carrying capacity were obtained from Israel, but they were received in a poor state of repair and the Rhodesians generally had maintenance problems with much of their equipment. The loss of a Bell Huey to a surface-to-air missile (SAM) inside Mozambique in September 1979, in which all 12 occupants were killed, was the single greatest disaster in terms of casualties suffered by the Rhodesians during the war.

The Fire Force concept represented what might be termed 'vertical envelopment' of the guerrillas and this technique was also utilised in external raids into Zambia and Mozambique in which the SAS and Selous Scouts often figured prominently. On other occasions, the Rhodesians drove (often in captured vehicles) or walked to their targets, while there were also more limited penetrations across frontiers by small groups of Rhodesians to lay mines or set up ambushes. External operations, however, not only tended to divert manpower from critical areas inside Rhodesia but also became more and more hazardous. Rhodesian command of the air was threatened by the deployment of SAMs in Mozambique, while the actual concentration of Rhodesian airpower in support of major incursions could itself give prior warning to the guerrillas. There was also a suspicion that the guerrillas were sometimes forewarned of Rhodesian operations by a source within the Security Forces at a high level. The main airstrike capacity, apart from the 13 or 14 Cessna Lynx converted to a counter-insurgency role, consisted of a squadron of Hawker Hunters. There was also a squadron of Canberra bombers of which four were used in the bombing raid over Angola in February 1979.

External operations by land could prove equally vulnerable as the guerrilla bases and camps, particularly in Mozambique, became ever better protected with sophisticated defences. The Rhodesian attack on New Chimoio in September 1979 is a case in point. A force of 100 Selous Scouts battled for three days successfully to overcome some 6000 ZANLA guerrillas and their East German advisers. Although the guerrillas eventually broke and fled, leaving over 3000 dead, they had had the benefit of an extensive trench and bunker system ringed by anti-aircraft, mortar and recoilless rifle positions. Many such bunker systems were largely immune to the ageing British 25-pounder field guns which comprised the bulk of Rhodesia's artillery.

The nature of the war inside Rhodesia also led to the development of a number of other special techniques by the Security Forces, the guerrilla penchant for attacking rural buses or civilian vehicles leading to the use of 'Q' cars — heavily armoured and armed decoy vehicles disguised as civilian traffic. The guerrillas' use of mines also led to the development of a large number of specially designed vehicles such as the Rhino, Hyena, Pookie and Hippo, which all featured a V-shaped body to deflect blast. Other trucks were sandbagged, while there was official promotion of a campaign to encourage driving at low speed to minimise the effectiveness of mines.

Inevitably, the guerrillas adjusted to Rhodesian tactics, often proving successful at exposing the Security Forces' observation posts[10] which were utilised to watch native villages for guerrilla presence. The kill ratio was invariably favourable to the Security Forces and never dropped below 6 to 1. At times it was as high as 12 or 14 to 1 overall, while individual operations might result in spectacular results of up to 60 to 1. The problem was that there was not the manpower to prevent increasing infiltration of Rhodesia. By the Security Forces' own estimates, the number of guerrillas operating inside Rhodesia grew from 350 or 400 in July 1974 to 700 by March 1976, 2350 by April 1977, 5598 by November 1977, 6456 by March 1978, to 11,183 by January 1979 and as many as 12,500 by the end of the war. The escalation of the conflict was also indicated by the expansion of JOCs from one to seven.

A number of options were available to try and control the extent of infiltration other than by spoiling attacks into host countries. One such method was the *cordon sanitaire* of border minefields established from May 1974 onwards. At a cost of 27,000 Rhodesian dollars per kilometre, a distance of 179 km (111 miles) between the Musengedzi and Mazoe rivers on the Mozambique frontier was fitted with a line of two game fences enclosing minefields and an alarm system. The system lacked depth, the tripwires and even the mines often being exposed by rainfall, and there were not sufficient numbers of men available for regular patrols along the fences. Later versions were widened, but enormous difficulties were experienced in maintaining the minefields. The lack of fencing alone resulted in a 30 per cent rate of replacement due to wild animals setting off the mines, while it was discovered that the guerrillas often removed claymore mines and used them as a 'topping' on their own land mines. In all some 864 km (537 miles) were eventually covered by the *cordon sanitaire* along the Zambian and Mozambique frontiers at a total cost of 2298 million dollars, but it remained only an impediment to infiltration and not an impassable barrier. In the process it consumed valuable resources that might have been utilised more effectively elsewhere.[11] By contrast the Botswana frontier was simply declared a free-fire zone.

The aim of preventing infiltration is to ensure that the guerrillas are separated from the civil population. Another common means of ensuring such separation since 1945 has been by resettling the population in protected areas. In Rhodesia resettlement was also utilised, many members of the Security Forces having witnessed its apparent success in Malaya. The initial project arose out of the extension of the 'no-go' area declared along the north-eastern frontier when insurgency mushroomed in late 1972, although a pilot scheme was first tried in the Zambezi valley in May 1973 whereby some 8000 Africans were resettled by December 1973. The main scheme then commenced with Operation 'Overload' in July 1974, by which over 46,000 Africans were removed from the Chiweshe TTL into 21 PVs and some 13,500 people from the Madziwa TTL a few weeks later. Resettlement was extended to areas not directly threatened by guerrillas in June 1975 with the creation of 'consolidated' villages or groups of *kraals* lacking the more direct protection afforded or theoretically afforded by PVs, but this was less successful and was dropped in 1976. Official figures indicate that there were 116 PVs by August 1976, 178 by September 1977 and 234 planned or built by January 1978. Estimates of the total population of PVs range from 350,000 to 750,000 Africans. Too frequently, however, PVs were regarded purely as a means of population control rather than as a basis for winning 'hearts and minds'. The fact that the scheme had begun in subverted areas rather than areas where the administration was sure of African loyalty was in itself an indication of the underlying motivation. Conditions naturally varied in PVs but too many lacked proper facilities and sanitation and it has been alleged that families were allocated as little as 12.5 sq. metres (15 sq. yards) each.[12] The villages were also inadequately defended with poorer quality Guard Force or DSAs, who often turned a blind eye to food being smuggled out to the guerrillas by a population which was, in any case, often insufficiently screened. Urbanisation also struck at the root of tribal values, especially among the Shona, as did restrictions such as dawn-to-dusk curfews, while crops cultivated at some distance from PVs were left unprotected by night and subject to animal depredations. Too often the Security Forces had forcibly removed the Africans to PVs and it is a measure of the failure of resettlement in Rhodesia that some 70 PVs in areas such as Mtoko, Mrewa and Mudsi had all restrictions lifted in September 1978 in the wake of the internal settlement. In almost every case the security situation immediately deteriorated, indicating how far the authorities had failed to win over the population.

Given the punitive nature of resettlement, it is perhaps little wonder that the winning of hearts and minds left much to be desired. An idea for a comprehensive scheme to win the loyalty of the African was in fact developed by Lieutenant Ian Sheppard in late 1973, the so-called 'Sheppard Group' of six men with marketing or public relations experience aiming to 'sell' the PVs to the Africans. Sheppard and his colleagues suggested some 38 different projects, including the establishment of an African Development Bank and granting land titles to resettled natives. Some suggestions were heeded, such as successfully persuading the Security Forces to innoculate native cattle against disease in the Masoso and Chinanda TTLs rather than slaughtering them wholesale. The majority fell foul of opposition from the Ministries of Internal Affairs and Information and the group folded in November 1974.[13] It was not until July 1977 that a Psychological Operations Unit was established under the direction of Tony Datton, a former member of the Sheppard Group, but continued rivalry with the Special Branch and resistance from senior military officers thwarted Datton's efforts. A Directorate of Psychological Warfare was belatedly established in 1979 but proved ineffectual. Similarly, Operation 'Manila Interface', initiated in August 1978 psychologically to prepare the ground for resettlement, was a failure.

Rather than attempting to provide the rural African with more facilities, there was a tendency to concentrate on broadening the representation of the African in government, but this meant little to the average African and rarely offered a viable alternative to guerrilla intimidation. Moreover, with the Security Forces intent on eliminating guerrillas rather than winning hearts and minds, the latter tended to consist of a 'carrot and stick' approach. Thus rewards for information ranging from 300 to 1000 dollars were introduced in April 1974 and these were backed by an extensive aerial propaganda campaign, dropping leaflets and safe conduct passes to guerrillas who might be willing to surrender. Full-scale amnesties were offered in both December 1977 and March 1979, with only limited success. The reverse of the carrot for co-operation was restriction and punishment. Collective fines were introduced in January 1973 if the presence of guerrillas was not reported within 72 hours, the fine being extracted in the form of livestock or, as in the case of Chiweshe TTL, in the form of enforced closure of African grinding mills and stores. Death sentences were introduced for harbouring guerrillas in September 1973 and the two pieces of legislation providing the legal basis for the enforcement of anti-terrorist measures — the Emergency Powers Act and the Law and Order Maintenance Act — were constantly updated. The former was amended 32 times and the latter 12 times between 1965 and 1977.[14] From January 1977 Operation 'Turkey' applied rationing to Africans residing in labourers' compounds at white-owned farms as a measure of food control. There was also the registration card or *situpa* for Africans, but this was of little use to the Security Forces as it contained neither photograph, description nor fingerprints of the holder.

A more successful aspect of Rhodesian psychological warfare was the contest with the guerrillas for the control of traditional spirit mediums among the Shona. A register of all such mediums was compiled at an early stage. While, for example, the guerrillas abducted a woman claiming to be the legs of the Nehanda spirit in November 1972, the authorities controlled a number of others who claimed to be the head. District Commissioners also used psychological tactics such as demonstrating their 'power' over tame wild animals, but it can be noted that these officials had wide discretion to conscript native labour and to inflict corporal punishment. The preference for control rather than concessions was also illustrated by the extension in the use of martial law to govern, its application increasing over some 70 per cent of the country by September 1978 and to over 90 per cent by September 1979. Yet further evidence of disregard for the African could be drawn, too, from the forcible eviction of the Tangwena tribe from their traditional homes in the Inyanga area under the provisions of the Land Tenure Act in 1969.

Nevertheless, as already indicated, the Rhodesians were heavily dependent upon black servicemen and police who were forthcoming in sufficient numbers to maintain voluntary enlistment until 1979, at which point black conscription was introduced as a political measure by the transitional government. Equally noteworthy was the successful use of pseudo forces, by which members of the Security Forces as well as captured guerrillas 'turned' by the former, were utilised to infiltrate guerrilla organisations. A pilot scheme was attempted in October 1966 by Senior Assistant Commissioner Oppenheim of the BSAP CID and others including Lieutenant Alan Savoury, who had experience of working in the game parks, and Lieutenant 'Spike' Powell who had worked with British pseudo-gangs in Kenya during the Mau Mau emergency of the 1950s. But, since the guerrillas had such little support in the Zambezi valley and were so easily contained, there was neither scope nor use for pseudo operations. The idea was only revived with the escalation of the war by ZANLA in late 1972 and indeed the pseudo-gangs were always more successful in penetrating ZANLA than ZIPRA since the former had a much looser discipline. Superintendent Tommy Peterson deployed the first pseudo team in Bushu TTL in January 1973, the concept of 'frozen' areas in which the teams could work without being killed by the Security Forces being adopted from August 1973. From these beginnings developed Reid-Daly's Selous Scouts a combat tracker unit, either observing guerrillas and guiding other units to the attack or themselves increasingly adopting a hunter-killer role. Employing guerrilla defectors from the start, the Selous Scouts were sometimes required to call in airstrikes close to their own positions to avoid disclosing their true identities. Similarly, the Selous Scouts appear to have attacked PVs on occasions to prove their *bona fides* in the course of seeking to sow distrust within the insurgent groups. Pseudo operations were always dangerous and required a constant supply of new defectors in order to enable the Scouts to keep up-to-date on guerrilla internal security measures. Their reputation was somewhat mixed, many senior military and police officers doubting the merit of releasing captured insurgents who would otherwise have faced the full force of the law. The Selous Scouts attracted the nickname of 'armpits with eyeballs' through their generally unkempt appearance.[15]

Other than relying on the Security Forces, there was always the possibility of arming loyal Africans, but this reached no further than deploying Africa DSAs in the PVs. In 1978, however, money became available from Oman[16] which enabled a pilot scheme to be launched in Msana TTL in March by which 90 local Africans were formed into an Interim Guard Force. With the internal settlement, the opportunity was available for recruiting more blacks loyal to Muzorewa, Sithole and Chirau, and the SFAs were quickly established to take over security duties in TTLs. Known in Shona as *Pfumo reVanhu* and in Ndebele as *Umkonto wa Bantu* (both meaning 'Spear of the People'), the SFAs were in reality private armies attached to Muzorewa's UANC and Sithole's branch of ZANU. Allegedly guerrillas who had accepted the latest amnesty, the SFAs were primarily black conscripts or unemployed urban blacks given a hasty four-week crash course of training. Under Operation 'Favour' some 2000 SFAs were deployed in 80 TTLs by the end of 1978 and their strength grew to some 10,000 in the run-up to the internal election in the following year. By April 1979 they had responsibility for 22 frozen areas representing some 15 per cent of the country as a whole. In theory the SFAs gave the Security Forces the ability to hold outlying areas on a permanent basis for the first time, but the SFAs were ill-trained and poorly disciplined. The situation did not materially improve after the Army's Special Forces Headquarters took over responsibility for the SFAs in July 1979 and the Army as a whole had little faith in their abilities. Indeed, one group of SFAs loyal to Sithole had to be eliminated by the Security Forces in the Gokwe TTL in June 1979. At most their deployment enabled the Guard Force to be switched to railways and farms, but the brutality of the SFAs in the TTLs did little to enhance support for the Muzorewa government.

The conditions under which many Africans were living in the

TTLs by the end of the war raises the wider question of the effects of guerrilla conflict upon Rhodesia and its people. For black civilians the war was immensely disruptive. Many Africans had become refugees crossing to Zambia or Mozambique voluntarily. Still others had been forcibly abducted by the guerrillas — in July 1973, for example, 273 African school-children were abducted from the St Alberts mission although most were subsequently recovered by the Security Forces. All the main white urban areas such as Salisbury, Bulawayo and Umtali had substantial native refugee populations on their outskirts while, of course, many Africans had been forcibly relocated by the Security Forces in PVs. The Africans were caught in the real sense between intimidation by both Security Forces and guerrillas, the ratio of black civilian casualties caused by the protagonists running at 40:60 by 1978. In rural areas it is clear that local administration had often broken down by the end of the war. It was reported in May 1977 that over 22,000 Africans in the southeast were refusing to pay taxes. By the end of 1978 over 900 African primary and secondary schools had closed, leaving over 230,000 pupils without access to education. Over 90 rural hospitals and clinics had also closed and rural bus services had been cut by 50 per cent. African agriculture was severely depressed, the Ministry of Agriculture calculating that 550,000 head of native cattle would perish in the course of 1978 alone. It is believed that over a third of the native cattle herd died during the war while, with only 1500 out of 8000 cattle dips still in operation in 1979, diseases previously extinguished, such as anthrax and tsetse, were again rampant. Prior to the war African farmers had produced 70 per cent of Rhodesia's food requirements, but by 1977 this had already fallen to only 30 per cent.[17]

For the white population there were parallel strains. Not only was there the burden of conscription but also the economic cost of the war. By December 1978 the war was costing a million dollars a day, defence expenditure having risen by a staggering 610 per cent between 1971–2 and 1977–8. That on police had risen by 232 per cent during the same period, with expenditure on internal affairs and roads rising by 305 per cent and 257 per cent respectively. It has been argued that the war was relatively cheap, Rhodesia spending less in 1978 and 1979 than the sum spent on the annual administration of the University of Berkeley, California[18] but, of course, it did not appear to be so to those experiencing it. South Africa may have subsidised the Rhodesian war effort by as much as 50 per cent, but there were still fairly constant tax increases such as the 12½ per cent surcharge on income tax imposed in July 1978. Similarly, the property market was depressed and tourism declined by some 74 per cent between 1972 and 1978. Coupled with sanctions, the war saw a decline in Rhodesia's GNP amounting to 1.1 per cent in 1975, 3.4 per cent in 1976 and 6.9 per cent in 1977. The physical strain of the war also resulted in rises in alcoholism, illegitimacy and divorce among white Rhodesians.[19] For white farmers in particular, the war meant constant danger and a life at night of floodlights, wire and sandbags. In January 1973 insurance firms had pronounced themselves unwilling to compensate for guerrilla action, leading to a Terrorist Victims Relief Fund in February and a government Victims of Terrorism (Compensation) Bill in June 1973. Officially the war cost the deaths of 410 white civilians and 954 members of the Security Forces. A total of 691 black civilians are said to have died and 8250 guerrillas, but these figures are clearly understated and it is possible that the total deaths exceeded 30,000.[20]

At the end of the war the Rhodesian Security Forces had surrendered no city or major communications route and the BSAP had closed no police station, even along the exposed Mozambique frontier. The guerrillas had not succeeded in establishing any 'liberated zones', although clearly large parts of Rhodesia were being actively contested. The guerrillas, indeed, have been characterised as the 'worst' this century[21] in terms of their military effectiveness and expertise, the Rhodesians referring to a so-called 'K' factor (for Kaffir) in this regard. The guerrillas were, however, effective in political subversion and whether the situation could have been maintained by the Security Forces indefinitely is a moot point. At the time of the ceasefire an estimated 22,000 ZIPRA and 16,000 ZANLA guerrillas remained uncommitted outside the country, although not all were trained. Within Rhodesia, even with the dubious addition of the SFAs, the ratio of the Security Forces to the guerrillas and their supporters reached only 1:1.5.[22] Manpower had always been the problem, particularly as the Rhodesians had attempted for far too long to exert control everywhere rather than consolidating their grasp of key areas. Militarily, the war was not lost by the end of 1979 despite the frequent lack of co-ordination in command, control and intelligence. However, Rhodesia's resources were stretched dangerously thin while the general approach of the Security Forces to counter-insurgency was not conducive to establishing any enduring popular African support. Overall lay the interplay of dominating political considerations that eventually determined the outcome. The legacy of the war was a newly-independent state beset by economic and social problems, not least the rivalries of the nationalists that the war had stimulated and left unresolved.

Notes

1. L.H. Gann and T.H. Henriksen, *The Struggle for Zimbabwe: Battle in the Bush* (Praeger, New York, 1981), pp. 81–2.

2. P.L. Moorcraft and P. McLaughlin, *Chimurenga: The War in Rhodesia, 1965–1980* (Sygma/Collins, Marshalltown, 1982), p. 36.

3. J.K. Cilliers, 'A Critique on Selected Aspects of the Rhodesian Security Forces' Counter-Insurgency Strategy, 1972–1980' (Unpublished MA, University of South Africa, 1982), p. 272.

4. For accounts of the early campaigns between 1966 and 1970 see J. Bowyer Bell, 'The Frustration of Insurgency: The Rhodesian Example in the Sixties', *Military Affairs* 35/1, 1971, pp. 1–5; M. Morris, *Terrorism* (Howard Timmins, Cape Town, 1971); K. Maxey, *The Fight for Zimbabwe* (Rex Collings, London. 1975); A.R. Wilkinson, *Insurgency in Rhodesia, 1957-1973* (International Institute for Strategic Studies, Adelphi Paper no. 100, London, 1973).

5. R. Reid-Daly and P. Stiff, *Selous Scouts: Top Secret War* (Galago, Albertown, 1982), pp. 260–74.

6. Moorcraft and McLaughlin, *Chimurenga*, p. 191.

7. Reid-Daly and Stiff, *Selous Scouts*, p. 68–9; Tony Geraghty, *Who Dares Wins* (2nd edn, Arms and Armour Press, London, 1983), p.296.

8. Cilliers, 'Critique', pp. 308–9.

9. Moorcraft and McLaughlin, *Chimurenga*, pp. 66–7.

10. T. Arbuckle, 'Rhodesian Bush War Strategies and Tactics: An Assessment', *Journal of the Royal United Services Institute* 124/4, 1979, pp. 27–32.

11. Cilliers, 'Critique', pp. 165–6.

12. T.J.B. Jokonya, 'The Effects of the War on the Rural Population of Zimbabwe', *Journal of Southern African Affairs*, 5/2, 1980, pp. 133–47; R. Marston, 'Resettlement as a Counter-revolutionary Technique' *Journal of the Royal United Services Institute*, 124/4, 1979, pp. 46–9.

13. Cilliers, 'Critique', p. 200.

14. Jokonya, 'Effects', pp. 133–47.

15. Reid-Daly and Stiff, *Selous Scouts*, p. 245.

16. Cilliers, 'Critique', p. 278.

17. Marston, 'Resettlement', pp. 46–9.

18. Gann and Henriksen, *Struggle for Zimbabwe*, p. 72.

19. Moorcraft and McLaughlin, *Chimurenga*, p. 174.

20. Ibid., p. 222.

21. N. Downie, 'Rhodesia: A Study in Military Incompetence', *Defence*, 10/5, 1979, pp. 342-5.

22. Cilliers, 'Critique', p. 296.

References

Arbuckle, T. 'Rhodesian Bush War Strategies and Tactics: An Assessment', *Journal of the Royal United Services Institute*, 124/4, 1979, pp. 27–32

Barclay, G. St J. 'Slotting the Floppies: The Rhodesian Response to Sanctions and Insurgency, 1974–1977', *Australian Journal of Defence Studies*, 1/2, 1977, pp. 110–20

Blake, R. *A History of Rhodesia* (Eyre Methuen, London, 1977)

Bowyer Bell, J. 'The Frustration of Insurgency: The Rhodesian Example in the Sixties', *Military Affairs*, 35/1, 1971, pp. 1–5

Bruton, J.K. 'Counter-insurgency in Rhodesia', *Military Review*, 59/3, 1979, pp. 26–39

Cilliers, J.K. 'A Critique on Selected Aspects of the Rhodesian Security Forces' Counter-Insurgency Strategy, 1972-1980' (Unpublished MA thesis, University of South Africa, 1982)

Cohen, B. 'The War in Rhodesia: A Dissenter's View', *African Affairs*, 76/305, 1977, pp. 483–94

Downie, N. 'Rhodesia: A Study in Military Incompetence', *Defence*, 10/5, 1979, pp. 342–5

Gann, L.H. and Henriksen, T.H. *The Struggle for Zimbabwe: Battle in the Bush* (Praeger, New York, 1981)

Good, R.C. *UDI: The International Politics of the Rhodesian Rebellion* (Faber and Faber, London, 1973)

Jokonya, T.J.B. 'The Effects of the War on the Rural Population of Zimbabwe', *Journal of Southern African Affairs*, 5/2, 1980, pp. 133–47

Lovett, J. *Contact* (Khenty, Johannesburg, 1977)

Marston, R. 'Resettlement as a Counter-revolutionary Technique', *Journal of the Royal United Services Institute*, 124/4, 1979, pp. 46–9

Martin, D. and Johnson, P. *The Struggle for Zimbabwe* (Faber and Faber, London, 1981)

Maxey, K. *The Fight for Zimbabwe* (Rex Collings, London, 1975)

Meredith, M. *The Past is Another Country* (Andre Deutsch, London, 1979)

Moorcraft, P. *A Short Thousand Years* (Salisbury, 1979)

Moorcraft, P. *Contact II* (Sygma, Johannesburg, 1981)

Moorcraft, P. and McLaughlin, P. *Chimurenga: The War in Rhodesia, 1965-1980* (Sygma/Collins, Marshalltown, 1982)

Morris, M. *Terrorism* (Harold Timmins, Cape Town, 1971)

Morris, M. *Armed Conflict in Southern Africa* (Cape Town, 1974)

Ranger, T. 'The Death of Chaminuka: Spirit Mediums, Nationalism and the Guerrilla War in Zimbabwe', *African Affairs*, 81/324, 1982, pp. 349–69

Reid-Daly, R. and Stiff, P. *Selous Scouts: Top Secret War* (Galago, Albertown, 1982)

Sobel, L.A. (ed.) *Rhodesia, 1971–1977* (Checkmark, New York, 1978)

Venter, A.J. *The Zambezi Salient* (Hale, London, 1975)

Wilkinson, A.R. *Insurgency in Rhodesia, 1957–1973* (International Institute for Strategic Studies, London, Adelphi Paper no. 100, 1973)

Wilkinson, A.R. 'From Rhodesia to Zimbabwe' in Davidson, B., Slovo, J. and Wilkinson, A.R. (ed.), *Southern Africa* (Penguin Books Ltd., Harmondsworth, 1976), pp. 215-340

Wilkinson, A.R. 'Introduction' and 'Conclusion' in Raeburn, M. (ed.), *Black Fire* (Julian Friedmann, London, 1978), pp. 1–52 and 233–43

Wilkinson, A.R. 'The Impact of the War' in Morris-Jones, W.H. and Austin, D. (ed.), *From Rhodesia to Zimbabwe: Behind and Beyond Lancaster House*

(Frank Cass, London, 1980), pp. 110–23. (This was previously published as Wilkinson, A.R. 'The Impact of the War', *Journal of Commonwealth and Comparative Politics*, 18/1, 1980, pp. 110–23)

7 THE SOUTH AFRICAN ARMY: THE CAMPAIGN IN SOUTH WEST AFRICA/NAMIBIA SINCE 1966

Francis Toase

South African para-military and military forces have been engaged in continuous counter-insurgency (COIN) operations in South West Africa/Namibia[1] since August 1966. This South African campaign, one of the longest but most obscure and under-reported COIN campaigns since the Second World War, can be divided into two phases. Initially, during the late 1960s and early 1970s, the campaign was very much a low-key affair. Insurgent activities were rather desultory and the South African Government attempted to counter the insurgents by deploying para-military policemen in the territory. Later, as the insurgents intensified their activities, Pretoria modified its policies. Primary responsibility for countering the insurgents was handed over to the Army and a comprehensive COIN strategy, that gave emphasis to the political and socio-economic as well as military aspects of COIN warfare, was introduced. This evolution of South African COIN strategy in South West Africa will be examined below in some detail. First of all, however, it is proposed to consider briefly the COIN campaign waged by South Africa against indigenous nationalist insurgents between 1961 and 1964, since the methods adopted in this campaign influenced Pretoria's initial approach in South West Africa. Passing reference will also be made to South African military involvement in Rhodesia, an experience that was also to influence South African strategy in South West Africa.

The immediate origins of the insurgency that occurred in South Africa during the early 1960s can be traced back to the period between 1948 and 1960, years of growing confrontation between the South African Government and the country's African nationalists. As the Government began to implement its policy of apartheid, strengthening white supremacy and institutionalising racial segregation, the African nationalists became increasingly militant. The African National Congress (ANC), which had previously sought to gain limited advances through constitutional

action, turned to unconstitutional methods such as civil disobedience in an effort to make the Government abandon apartheid
and move towards a multiracial, democratic, socialist system. The
Pan-Africanist Congress (PAC), a rival nationalist movement
formed in April 1959 by former ANC members, also adopted
unconstitutional methods, its objective being a system of
government of, by and for Africans only. However, neither
nationalist movement deflected the Government from its chosen
course. As far as the Government was concerned, African
majority rule, of either the ANC or PAC variety, was anathema
and the nationalists were threatening the security of the State.
Indeed, the Government suppressed the African nationalist
protests with increasing severity during the 1950s and eventually,
in the wake of the Sharpeville shootings of 21 March 1960,
outlawed both the ANC and the PAC.

The African nationalists responded forcefully. Concluding that
non-violent protest had failed, both the ANC and the PAC
formed clandestine military wings with a view to overthrowing
the white South African State by force of arms. The ANC, in
conjunction with the outlawed South African Communist Party
(SACP), created *Umkhonto we Sizwe* ('Spear of the Nation'), a
predominantly, though by no means exclusively, black African
movement. PAC members founded *Poqo*, whose name — 'only'
or 'pure' — indicated that it was an exclusively black African
movement. *Umkhonto we Sizwe* decided to wage a sabotage
campaign, so as to disrupt communications, undermine order and
demoralise the authorities and their supporters; it also sent
recruits abroad for military training, so that at a later stage it
might initiate a guerrilla war in selected areas of the country.
Poqo opted for a terrorist campaign, to be directed against whites
and against blacks, Asians and Coloureds (persons of mixed race)
who supported the authorities. *Poqo*'s campaign began early in
1962 (no precise date has ever been given), while *Umkhonto*'s
began on 16 December 1961, when bomb attacks were carried
out against Government installations and buildings in Johannesburg, Port Elizabeth and elsewhere.

At the time, this twin insurgent threat appeared to pose a
serious challenge to the South African Government. The
authorities faced the task of securing a vast country,
1,221,037 sq.km (471,445 sq. miles) in size, containing 16 million
people — three million whites, one and a half million Coloureds,

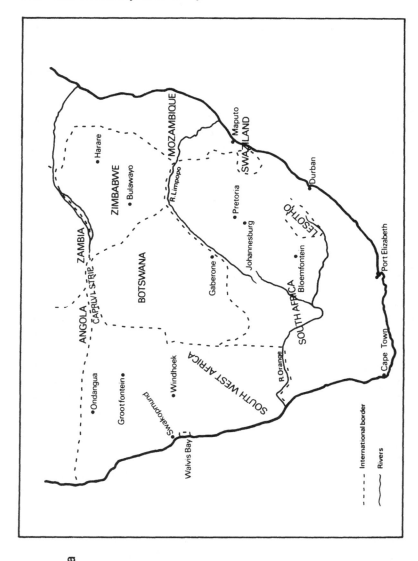

Map 7.1:
South West
Africa/Namibia

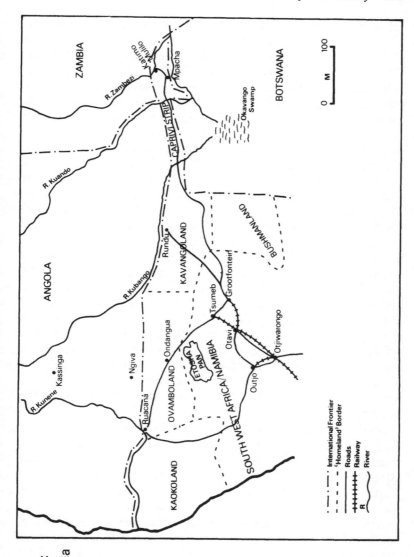

Map 7.2:
The War Zone:
Namibia/Angola

half a million Asians and eleven million black Africans or 'Bantu'
— and an extensive and vulnerable economic infrastructure. The
personnel available to perform this task were far from numerous.
The South African Defence Force (SADF), an all-white force
based on a small nucleus of regulars (the Permanent Force) and
selective national service, could muster some 78,000 men, if the
Citizen Force (former national servicemen available for further
periods of part-time service) and Commandos (local militias)
were included; its standing operational strength, however, was
only 11,500.[2] The South African Police (SAP) numbered some
26,000, of whom over 13,000 were black.[3] Moreover, neither the
SADF nor the SAP could boast of any recent experience in the
COIN sphere. The Air Force had taken part in the Berlin airlift
(1948–9) and the Korean War (1950–3), but the Army had seen
no active service between 1945, when it was demobilised after the
Second World War, and 1961, when it was put on a state of
'preparedness for service' to deal with internal unrest. The SAP
was in a better position, for having been the State's first line of
internal security since 1948, it was well qualified to deal with civil
disobedience. Even so, the SAP's most recent experience of
sabotage attacks was back in 1941–2, when an extremist
Afrikaner movement called the *OssewaBrandwag* ('Ox-Wagon
Sentinel') had used sabotage against the pro-British Smuts
regime. In many ways, therefore, the South African authorities
were not well prepared to deal with the African nationalist
insurgents.

On the other hand, the authorities did have certain advantages
over the insurgents. In the first place, the insurgents were
isolated from outside support. The ANC and the PAC had
attracted considerable international support, notably from the
United Nations (UN) and from the independent black African
states, some of which proposed to give military assistance to the
insurgents. However, between black Africa and the Republic of
South Africa lay a protective barrier of territories that were
either under South African administration, namely South West
Africa, or economically dependent upon the Republic, namely
Bechuanaland (later Botswana), Basutoland (later Lesotho) and
Swaziland, or under the control of regimes antipathetic towards
African nationalism, namely white-settler dominated Southern
Rhodesia (later Zimbabwe) and the Portuguese territories of
Angola and Mozambique. This *cordon sanitaire* rendered the

infiltration of men and arms into the Republic well-nigh impossible. Second, behind this defensive glacis the Security Forces and the white community possessed a virtual monopoly of arms, the only non-whites who had official access to arms being certain black members of the SAP. Third, insurgent activities were hampered by the state's apartheid regulations. The African homelands or 'Bantustans' were governed by chiefs and officials in the employ and pay of the South African Government, while movement of Africans in so-called white areas was rigidly controlled by the 'influx' and 'pass' laws, and Africans who lived in white urban areas were accommodated in separate townships easily sealed off by the Police. Fourth, the authorities began with a head start over the insurgents in terms of intelligence. The insurgents had few informers within the state apparatus, whereas the Government, or more specifically the SAP, had a network of informers among all the races, including the Africans. In any case, many of the leading members of the nationalist underground movements were known to the authorities from the days of open protest between 1948 and 1961.

This intelligence advantage proved to be invaluable, since the Government's campaign against the insurgents was based not on military firepower but on standard Police operations designed to apprehend persons collectively referred to as terrorists and criminals. Indeed, the Army was used as an adjunct to the Police, rather than vice versa. The Army was enlarged considerably[4] and troops were used to guard key installations such as dams, power stations and oil storage depots. However, the primary role in the campaign fell to the SAP, and in particular to the Security Police (formerly known as the Special Branch). Under B.J. Vorster, who became Minister of Justice in August 1961, the Security Police branch was strengthened and the SAP in general enlarged to 34,000 men by 1964;[5] this strength was augmented by the establishment in 1961 of a Reserve Police force, whose civilian volunteers could be called upon to perform ordinary police duties, thus releasing regulars for more urgent duties. Under Vorster's tenureship the SAP was also given greatly extended powers of arrest and detention, designed to help the Police uncover plots and obtain information from detainees.

Indeed, the work of the Police was greatly facilitated by a series of draconian security laws enacted by the South African Government after 1961. Existing security laws like the Suppression of Communism Act of 1950[6] and the Public Safety Act of 1953[7] were supplemented by new legislation such as the General Laws Amendment Acts of 1961, 1962, 1963 and 1965. Sabotage was made a statutory offence, was given a very broad definition and carried severe penalties: convicted saboteurs were liable to a minimum sentence of five years imprisonment and a maximum sentence of death. Moreover, the Police were given the authority to detain suspects without charge, and under conditions of solitary confinement, for twelve days (1962), 90 days (1963) and 180 days (1965). The power of the State to hold suspects incommunicado, together with earlier legislation that had made the unauthorised reporting of prison conditions illegal, meant that the Police were free to use whatever methods they saw fit to obtain information from suspects. This series of draconian regulations, willingly accepted by the (overwhelmingly white) electorate in order to preserve the *status quo*, greatly reduced the rights of the individual *vis-à-vis* those of the State and gave the latter an enormous advantage in its campaign against the insurgents.

It did not take the Police long to produce the intended results. During 1962 and early 1963 *Poqo* had *inter alia* fomented riots, attacked power plants, railway lines and police stations, and murdered several African chiefs and white civilians. By May 1963, though, over 2000 *Poqo* suspects had been detained and the movement all but broken. *Umkhonto we Sizwe*, which carried out most of the 203 serious cases of sabotage reported by the Police up to 10 March 1964, did not last much longer than *Poqo*. The Police had arrested *Umkhonto's* leader, Nelson Mandela, in August 1962. The *coup de grace* was delivered on 11 July 1963, when the Police executed a well-prepared raid on the ANC/SACP High Command's supposedly secret headquarters at Lilliesleaf Farm, Rivonia, mid-way between Pretoria and Johannesburg. The SAP seized 17 persons, constituting virtually the entire leadership of the High Command, plus valuable files and documents. Another insurgent group, the African Resistance Movement (ARM), suffered the same fate as *Poqo* and *Umkhonto we Sizwe*. Predominantly white in membership, the ARM was responsible for most of the sabotage acts carried out in the first half of 1964, but by July/August of that year most of the movement's saboteurs had been arrested. By the end of 1964, few insurgents of any political denomination remained at large within South Africa, those who had managed to evade detention having fled abroad. By early 1965 sabotage had virtually ceased. The insurgent campaigns had failed.

This failure can be attributed in some measure to the insurgents' own mistakes and miscalculations. For one thing, the insurgents appear to have given insufficient attention to practical issues such as the availability of aramaments. Infiltrating arms into the Republic or seizing them from the Security Forces proved to be extremely difficult. *Poqo* had to fall back upon the use of knives, axes, pangas and assegais, while *Umkhonto*'s home-made bombs were often of poor quality. Second, the insurgents misjudged the political mood of the non-white population. Both the ANC and the PAC expected, or at least hoped for, a popular uprising, but in the event these hopes were dashed. The black African population refused to respond, either because they did not support the insurgents, or because they feared the consequences of opposing the authorities, or both; the PAC compounded its difficulties by adopting the slogan *Poqo*, an emotional we-alone Africanist cry that alienated potential sympathisers within the Asian, Coloured and white communities. The insurgents also misjudged the white ruling caste. The Verwoerd Government was neither weak nor corrupt. It was tough, uncompromising, ruthless and was determined, despite intense international hostility, to preserve white supremacy in South Africa.

Indeed, the insurgents' failure was also due, in no small measure, to the stand taken by the Government. The latter did not make any political concessions to the nationalists, though it can be argued that by accelerating the 'Bantustan' programme (that is, the policy of granting self-government to the African 'nations' in their own territorial homelands) the Verwoerd regime went some way towards undercutting revolutionary African nationalism. Nevertheless, the Government did find an effective way of countering the insurgents: administrative repression coupled with police action. Parliament enacted the necessary legislation and the SAP, which the ANC/SACP underestimated to their cost, acted with ruthless efficiency. Thus were the insurgent groups smashed and the possibilities of a guerrilla uprising pre-empted.

The suppression of *Poqo* and *Umkhonto we Sizwe* earned the Republic of South Africa a lengthy respite from insurgent attack.

The African nationalists regrouped outside South Africa and made several attempts, in 1967 and 1968, to infiltrate guerrillas into the Republic[8], but these efforts proved abortive. The guerrillas were frustrated by the *cordon sanitaire* between black Africa and the target state. Not until 1977, two years after the advent of a black African, Marxist regime in Mozambique, did the nationalists resume their attacks on the Republic. And not until the early 1980s, when the ANC/SACP attacked several prestige targets, did the insurgents really make their presence felt.

This is not to suggest, however, that the South African Security Forces carried out no COIN operations between 1964 and the early 1980s. The threat to the Republic itself may have receded by early 1965, but the South African Government became increasingly concerned about the insurgent threats to the nearby territories of Angola, Mozambique and Rhodesia. South Africa contributed to the COIN efforts waged in all three territories. In addition, she fought a major COIN campaign of her own in a fourth buffer territory, namely South West Africa, from 1966 onwards.

South Africa's contribution to the Portuguese campaigns in Angola and Mozambique appears to have been very limited. Pretoria has never actually admitted to making any contribution at all, but even if allegations to the contrary are accepted, her involvement was not particularly heavy. It has been suggested that around 1000 South African troops helped the Portuguese in Mozambique,[9] while in Angola a joint Portuguese-South African command centre was set up in the late 1960s to direct South African air reconnaissance and troop transport activities against Angolan and South West African insurgents.[10]

Pretoria's involvement in Rhodesia was more extensive. South African forces were first deployed in Rhodesia in August 1967, ostensibly to intercept a group of (South African) ANC insurgents who were *en route* from Zambia to the Transvaal. The guerrillas were decimated by Rhodesian and South African Security Forces, as were other ANC guerrillas who infiltrated into Rhodesia in 1968. Be that as it may, South African forces — officially described as para-military policemen trained in COIN techniques — remained in Rhodesia until August 1975. Their average strength was about 2000,[12] that is, nearly half the entire strength of the regular Rhodesian Army at that time. Even after

Pretoria officially withdrew its 'policemen' in August 1975, South African involvement in Rhodesia continued. Pretoria supplied much of the equipment used by the Rhodesians and individual South Africans were recruited into the Rhodesian forces in considerable numbers.[12] Moreover, towards the end of the war in the late 1970s, the South African Government began to take a more active military role in order to prop up the Muzorewa regime. It has been reported that pilots, engineers and gunners were seconded to the Rhodesian forces and that as many as 6000 SADF personnel were stationed in Rhodesia by early 1979, though this figure may well be an exaggeration. Officially Pretoria had only 400 or so troops in the country at that time, with the limited function of guarding the rail/road border crossing at Beitbridge.[13] Whichever version is accepted, the fact remains that South Africans did serve in Rhodesia between 1967 and 1980, either in SAP or SADF uniforms or in an individual capacity within the Rhodesian armed forces. They thereby picked up valuable experience of COIN techniques, including the Rhodesians' successful development of so-called Fire Force operations and cross-border raids on insurgent camps.[14]

Angola, Mozambique and Rhodesia were not the only buffer territories in which South African forces participated in COIN operations during the late 1960s and 1970s. The SAP and SADF were also involved in operations in South West Africa, an adjacent territory rich in minerals, vast in size — 824,269 sq.km (318,252 sq miles) — but small in population, containing scarcely more than half a million people by the mid-1960s. In this particular case, though, the role of the South African forces was not that of lending support to another government's COIN effort. Rather, the South Africans fought their own COIN campaign in a territory under the *de facto* administration of the South African Government.

South West Africa had first come under South African control during the First World War, when in 1915, at Britain's behest, South African forces had invaded and seized the then German colony of *Süd West Afrika*. The territory was subsequently administered as an integral part of South Africa under a 'C-class' mandate of the League of Nations. After the demise of the League, in 1946, Pretoria took the process of integration a stage further, giving the territory's white population (some 14 per cent of the entire population) representation in the South African parliament in 1949 and placing the territory's 'Bantu' population under the authority of the South African Native Affairs Department in 1955. Pretoria also proceeded to apply its policy of apartheid in South West Africa. Racial segregation was strengthened and in 1964 Pretoria endorsed the findings of the Odendaal Commission, which had been set up in 1962 to make recommendations as to the political future of the territory. The Commission ruled out the formation of one central authority based on universal suffrage, recommending instead that each of South West Africa's eleven main ethnic groups be given self-government within its own separate area. Most of the Police Zone (the southern two-thirds of the territory) was to remain a white area, while the northern parts of the country, plus parts of the Police Zone, were to be reserved for the ten non-white groups.

The South African Government's control over South West Africa did not go unchallenged, however. The League of Nations' successor, the UN, had rejected Pretoria's claim to the territory in 1946 and during the next decade continuously urged Pretoria to place the territory under the UN's International Trusteeship System, pending full independence. By the early 1960s the UN's attitude had hardened. The General Assembly, reflecting the views of its ascendant Afro-Asian majority, began to demand independence for South West Africa. Notwithstanding the fact that the International Court of Justice had delivered a non-committal legal opinion in July 1966, the General Assembly proceeded in October of that year to revoke South Africa's mandate over South West Africa. In June 1968, the Assembly then renamed the territory Namibia and called for its independence. Three years later, in July 1971, the International Court declared South Africa's presence in 'Namibia' to be illegal.

In the meantime, South Africa's presence in South West Africa had come under challenge from within the territory, notably from newly-formed nationalist movements. The first such movement to be established was the South West African National Union (SWANU), in May 1959. However, SWANU's support was limited mostly to the Herrero tribe, a comparatively small ethnic group. By contrast, a rival nationalist movement established in April 1960, the leftist South West Africa People's Organisation (SWAPO), represented a major political threat to Pretoria's authority. SWAPO was based on the populous Ovambo tribe (the Organisation had begun life in 1957 as the Ovamboland People's Congress, renamed the Ovamboland People's Organisation in 1958), whose 270,000 people constituted the largest single ethnic group in the country, some 46 per cent of the entire population. Moreover, SWAPO represented a military as well as a political threat to the South African Government, because whereas SWANU had declined to take up arms, SWAPO decided to use force in order to oust the South Africans from South West Africa.

SWAPO's armed struggle was organised initially from Tanzania. In 1961 Sam Nujoma and other SWAPO leaders had fled to Dar-es-Salaam, where they set up headquarters. South West Africans were secretly recruited for guerrilla training, and after receiving such training in Tanzania, Algeria, Egypt, the USSR and elsewhere, returned to South West Africa via camps in Tanzania and Zambia. The first batch of insurgents infiltrated into South West Africa in late 1965. Equipped with Soviet and Chinese weapons, they proceeded to establish base camps in the centre-north of the country — Ovamboland —with a view to training more recruits *in situ*. One of these camps, located at Ongulumbashe (or Umgulumbashe), was discovered by the SAP and on 26 August 1966 was attacked. Two guerrillas were killed and eight captured in a clash that was subsequently honoured by SWAPO as marking the opening of the 'final phase' of its 'liberation struggle'.

At the time, SWAPO's chances of success were somewhat limited. Few trained insurgents were available, SWAPO's fighting strength being numbered in hundreds rather than thousands.[15] Moreover, most of South West Africa's huge land area was arid, thus lending itself to South African aerial reconnaissance. SWAPO's main target area, Ovamboland, was an exception to this rule in that parts of it were thickly wooded and other parts were covered in abundant vegetation during the rainy season (the summer months of December to March). However, access was a problem. Ovamboland was situated a considerable distance away from SWAPO's bases in Zambia, and the insurgents would have to infiltrate either through the south-eastern part of Angola, which was under Portuguese control, or through the north-eastern part of South West Africa, where the SADF had a number of military bases. Even if the insurgents did manage to reach Ovamboland, they would run a considerable risk of being killed or captured, since the tribal authorities were by and large loyal to the South African Government.

Undaunted by such considerations, SWAPO persevered. During the next eight years a pattern of guerrilla activity was established. Small groups of insurgents, usually armed with light weapons of Eastern bloc manufacture, sought to infiltrate South West Africa with a view to conducting sabotage attacks against installations, killing pro-Government blacks and harassing the Security Forces; southern Angola was sometimes used as an infiltration route but more often than not the insurgents used the Caprivi Strip, a 402 km. (250 mile) long panhandle located between Angola, Zambia and Botswana. At the same time SWAPO's in-country adherents sought to politicise the people, mobilise support and establish a viable political structure within South West Africa.

The task of countering these SWAPO activities fell primarily to the SAP, with the SADF playing a supporting role. This decision to give the SAP primary reponsibility for countering the insurgents was motivated partly by political considerations: the South African Government was anxious to portray the conflict as one between policemen and criminals rather than as one between rival groups of combatants. But the decision can also be attributed to practical considerations. The SAP, or at least certain elements thereof, had received specialised COIN training, whereas the Army had not. Moreover, as in South Africa itself between 1961 and 1964, intelligence gathering was considered to be a vital part of the COIN effort. In this respect too the SAP had the edge over the Army.

The SAP soon established a pattern to counter SWAPO's operations. Up-country, the SAP attempted to prevent the guerrillas from gaining access to South West Africa by blocking off the infiltration routes through the Caprivi Strip. Armed policemen, equipped with two-way radios and armoured cars, conducted continuous patrols in the Strip in an effort to locate and track down insurgents; similar patrols were carried out in Kavangoland and Ovamboland. These operations involved perhaps as many as 3000[16] men and were supported by aeroplanes and helicopters based at Mpacha, Rundu, Ondangua and elsewhere. In-country, the Security Police monitored SWAPO's political activities and when it was deemed necessary detained known or suspected SWAPO adherents, though SWAPO's internal wing was never banned as such. Detentions were legalised by new legislation passed by the South African Government, such as the Terrorism Act of June 1967. The Terrorism Act was made retroactive to July 1962 (when recruits were first sent for guerrilla training) and authorised the Police to hold suspects incommunicado for indefinite periods. It also provided stringent penalties for 'terrorism', a crime very broadly defined, with punishment ranging from five years imprisonment to the death penalty. The Act was used to detain captured insurgents as well as those arrested in-country.

These COIN measures met with some success. In-country, the SAP managed to keep the nationalists off-balance by circumscribing any overt political activities and by detaining SWAPO adherents. Along the border, the activities of SWAPO's military wing, the People's Liberation Army of Namibia (PLAN), were kept in check. PLAN issued grandiose propaganda statements, but in reality its military impact appears to have been negligible. Intermittent attempts were made to infiltrate groups of insurgents through the Caprivi Strip, but these often failed. Nor did PLAN inflict many casualties on the Security Forces. According to the South Africans, the Security Forces lost only eleven dead to mid-1974, the majority of these being killed in landmine explosions.[17] Not surprisingly, therefore, Pretoria saw no reason to make concessions to either the nationalists or the UN. Instead, the South African Government pressed ahead with the Odendaal plan. In 1969 South West Africa was incorporated into the Republic as a *de facto* fifth province. Not long afterwards, in May 1973, two of the largest Bantu homelands, Ovamboland and Kavangoland, were declared to be self-governing territories.

It was in the political sphere, however, that the South African Government began to run into difficulties. Although Pretoria kept the military situation under control, its political policies produced a fair amount of dissension and SWAPO was able to exploit this to good effect. The first eruption of popular discontent had occurred on 13 December 1971, when Ovambo contract workers, who provided about 70 per cent of the labour used in the white areas, went on strike in an apparently spontaneous protest against the contract labour system. This action developed into a general strike, and although the strike itself was soon settled by negotiation between the Government and the Bantu authorities, unrest continued, especially in

Ovamboland. Border posts were torn down, property was destroyed and headmen and chiefs attacked. Furthermore, as a result of a SWAPO-inspired boycott, only 2.3 per cent of the electorate bothered to vote in Ovamboland's first homeland election in August 1973.

These developments came as something of a shock to the South African Government, particularly as the Ovambo had up until that time been regarded as a 'loyal' tribe. Pretoria's response was correspondingly severe. Police reinforcements were flown in, quasi-martial law was introduced in Ovamboland in February 1972, and the Ovambo tribal authorities were allowed to use corporal punishment, including public flogging, against dissidents. Pretoria also increased the SADF presence in South West Africa. As the SAP became overstretched, Army units were sent in to help maintain order. By June 1974 the Army had officially assumed responsibility for COIN operations along the border.[18] Thus success in the military sphere notwithstanding, Pretoria was compelled to take drastic measures during the early 1970s, as internal unrest supplanted infiltration as the chief threat to the maintenance of the *status quo* in South West Africa.

Over the next few years the security situation in the territory deteriorated even further. The overthrow of the Portuguese Government in April 1974 and the subsequent decision by the new regime to withdraw from empire, dealt a severe blow to Pretoria's hopes of containing the SWAPO threat. As the Portuguese pulled out of Angola, SWAPO began to set up bases in the southern part of that country, at first covertly and later, after the *Movimento Popular de Libertação de Angola* (MPLA) emerged victorious from the Angolan civil war, with the official blessing of the new regime in Luanda. This opened up South West Africa's 1609 km. (1000 mile) long northern border to SWAPO infiltration. Moreover, SWAPO became militarily stronger. Guerrilla recruits began to flow into southern Angola in increasing numbers, training and arms being provided *in situ* by Cuban, Soviet and other Eastern bloc personnel; it has been suggested that by mid-1976 SWAPO's military strength stood at 2000[19] and that by early 1978 numbers had risen to perhaps 10,000.[20] SWAPO also made advances in the international sphere. In December 1976 the UN General Assembly expressed its support for SWAPO's armed struggle. The Assembly also invited SWAPO, having already (in December 1973) recognised the movement as the legitimate representative of the Namibian people, to participate as an observer in certain UN activities. It was against this background that SWAPO intensified its armed struggle after 1974. Operations were extended from the Caprivi Strip to Kavangoland, to the white areas around Grootfontein-Tsumeb-Otavi, and in particular to Ovamboland, where the previous confrontation between SWAPO's political cadres and the SAP gave way to military conflict between PLAN and the SADF.

These developments forced the South African Government to reconsider its policies in South West Africa. During the first eight years of the conflict, Pretoria had been able to contain the SWAPO threat by deploying small numbers of SAP personnel in the territory. By 1974, though, such an approach was no longer tenable. As the balance of power shifted in favour of SWAPO, indeed, the South African Government began to modify its policies. It embarked on a massive military build-up in the territory and introduced a comprehensive COIN strategy that gave attention to all the dimensions of COIN warfare, rather than just to the military dimension. SWAPO's intensification of the war was thus matched by fundamental changes in South African policy.

At the political level, the South African Government responded with a series of initiatives designed to undermine SWAPO's political appeal and deflect international pressure. The first such initiative was launched within six months of the Portuguese *coup d'état*. Representatives of each of South West Africa's eleven ethnic groups were invited to attend a constitutional conference in order to devise a formula for the independence of their country. This conference was held at the Turnhalle in Windhoek and began in September 1975. Two major decisions resulted from the talks. One was that South West Africa should become an independent unitary state. The other related to the system of government to be adopted by the new state. It was agreed that South West Africa should have a three-tier system, namely, a local government tier, another tier corresponding to the homelands, and a national tier comprising a National Assembly and a Ministerial Council chosen by the Assembly. The Council was to consist of one representative from each ethnic group, with decisions to be taken by consensus — in other words, each group would have a veto.

This dispensation suited Pretoria well enough. The homelands scheme would be preserved, albeit in modified form, at the regional level of government, while the interests of the whites would be protected by a veto in the Council of Ministers. Nevertheless, the South African Government declined openly to endorse the conference's decisions. The Turnhalle dispensation failed to attract international acceptance and under strong pressure from the West, Pretoria agreed to suspend the conference and negotiate an internationally acceptable settlement with the UN. Accordingly, Pretoria made further political changes in the territory. White South West Africans lost their representation in the South African parliament and an Administrator-General, empowered to rule by decree, was appointed as an interim measure pending independence. The first such incumbent took up his appointment on 1 September 1977.

This did not mean, however, that the ideas formulated at the Turnhalle talks were completely discarded. In November 1977 delegates who had participated in the talks formed the Democratic Turnhalle Alliance (DTA), a conservative, multiracial party that wished to achieve independence under South African tutelage and military protection. This party attracted South African backing. Indeed, while keeping its options open and playing along with international negotiations, the South African Government attempted to build up the DTA as a viable political alternative to SWAPO. In this respect the ground was prepared by the Administrator-General, who promulgated a flurry of decrees easing the apartheid regulations and repealing certain discriminatory laws. Subsequently, in December 1978, the South African Government held internally-supervised elections, with a view to setting up a national government in South West Africa. These elections, boycotted by SWAPO and denounced by the UN, produced a DTA victory, the latter party winning 82 per cent of the votes cast. The DTA then went on to dominate the interim Constituent Assembly and its successor the National Assembly, which was allowed to repeal or amend previous (South African) legislation and to make new laws. Pretoria retained control over security, foreign relations and constitutional matters, but the day-to-day running of South West Africa was handed over to the Assembly and its Council of Ministers, set up in July 1980. In effect, therefore, the DTA was installed as the 'Government' of South West Africa. It was thus given the chance to enhance its political reputation and steal SWAPO's political thunder. To some extent it succeeded in doing so, though its reforms tended to be blocked at the regional and local levels of government by the vociferous white community. By January 1983, indeed, the DTA Council of Ministers had resigned, ostensibly because of the slow pace of reform. The South African Government then dissolved the National Assembly, but a political successor to the DTA did emerge in the form of the Multi Party Conference, a six-party alliance that included the remnants of the DTA.

While the South African Government took political initiatives in order to take support away from SWAPO, it also attempted to check SWAPO militarily in order to give its political initiatives a chance to work. This task of checking SWAPO militarily required a considerable build-up of men and equipment. The details of this build-up have not been made available, but various estimates have been offered. According to the (pro-SWAPO) International Defence and Aid Fund, for example, South African troop levels in South West Africa stood at 15,000 in June 1974, were raised to 45,000 by early 1976 and had reached 100,000 by the early 1980s.[21] By contrast, less partisan observers have suggested that South Africa had no more than 15,000–20,000 troops in the operational areas by the early 1980s, and that the figure for the whole of South West Africa was perhaps 30,000–40,000.[22] Whatever the true figures, it cannot be gainsaid that Pretoria devoted a large proportion of its military resources to South West Africa after 1974. SAP units were joined by thousands of Permanent Force, Citizen Force and National Service personnel, by Commandos (until 1983) and by non-white South Africans serving in units such as the Cape (Coloured) Corps and 21 Infantry Battalion.[23] Pretoria also deployed artillery, armour and various types of aircraft — including Pumas, Alouettes, Mirages, Impalas, Buccaneers and Canberras — in the territory, thus making rather a mockery of the arms embargo imposed by the UN Security Council in 1963.[24]

This build-up of South African military strength was accompanied by a policy of what might be termed 'Namibianisation'. Partly to complement its political initiatives, partly to transfer the military burden to South West Africans, Pretoria sought progressively to tap local manpower, both white and black. Whites had been eligible for military service in the SADF for many years

(South West Africa had been an integral part of the SADF defence structure since 1939), but after 1974 the SADF intensified its efforts to raise white manpower. In July 1974 the SADF began to organise local Commando units, modelled on South Africa's own Commando units. By August 1980 there were 26 such units, known as Area Force Units, in existence. More innovative was the decision to recruit and train non-white South West Africans. Prior to 1974, non-white South West Africans had not been deployed in a combat role, though San (Bushmen) personnel had served as trackers. After 1974 this policy was modified and Pretoria launched an intensive drive to recruit non-whites. Initially units were raised on an exclusively ethnic basis, army battalions being forced from the San (1974), Ovambo, Okavango and East Caprivi (1975) peoples. From 1977 on, the SADF also attempted to create multi-ethnic units. These multi-ethnic and tribal units, together with the Area Force Units, were grouped together in August 1980 to form the South West Africa Territory Force (SWATF). Ninety per cent of SWATF's personnel were black, and although most of its officers were whites, seconded from the SADF, Pretoria quickly took steps to rectify this imbalance. A leadership training school was established at Okahandja, north of Windhoek, with a view to creating a black officer corps. Measures were also taken to build up an autonomous multiracial police force in the territory. These measures culminated in the creation of the South West Africa Police (SWAP) in April 1981. Moves to create indigenous Security Forces were paralleled by the extension of conscription to males of all races in South West Africa in January 1981, though in practice exemptions were granted to ethnic groups in the north and national service was applied selectively elsewhere.

Despite the huge build-up of SADF forces and the progressive Namibianisation of the war, Pretoria continued to exercise tight control over the various Security Forces operating in South West Africa. Nevertheless, various modifications in the command and control sphere were necessary from time to time. Before 1974, military activities in South West Africa had been the responsibility of the SWA Command in Windhoek, one of nine territorial commands of the SADF, while police operations had been the responsibility of the Divisional Commissioner of the SAP in Windhoek, who was in turn responsible to the Commissioner of the SAP. Between 1974 and 1977 these arrangements were amended. The maintenance of 'law and order' in-country remained the responsibility of the SAP's Divisional Commissioner, but with the SADF's assumption of control over the border areas, COIN operations in those areas fell under the authority of 101 Task Force in Grootfontein, itself directly responsible to the headquarters of the SADF in Pretoria.

Further changes followed in 1977, after the South African Government had decided to establish a multiracial army and self-sufficient military infrastructure in South West Africa. From August 1977 all military activities in the territory, including those of 101 Task Force and of the separate command which had existed up until that time in the enclave of Walvis Bay, were brought under the wing of the Defence Headquarters in Windhoek. Through this command the SADF and in turn the South African Government exercised full control over all military personnel and all COIN activities in South West Africa, including the operations of the SAP in the operational areas (where they operated as adjuncts to the SADF). In theory, these arrangements were modified by the inauguration of SWATF and SWAP, since both of these forces were placed under the control of the central authorities in Windhoek. In practice, however, few real changes ensued. SWATF's authority did not extend to the operational areas (nor to Walvis Bay), which remained under the authority of the SADF, and the General Officer Commanding SWATF in any case doubled as the Commanding Officer of all SADF troops deployed in South West Africa. In effect, therefore, Pretoria continued to exercise full control over all the Security Forces in the territory. It was thus able to avoid many of the difficulties associated with divided or weak COIN command structures.

Be that as it may, countering SWAPO was a formidable task. Initially, the South African Government attempted to deliver a crushing blow to SWAPO by intervening in Angola. This intervention had begun in August 1975, when a small number of SADF troops were sent over the border to protect the Ruacana dam from attack by SWAPO/MPLA forces. Later, encouraged by certain Western and black African States, Pretoria escalated its operation, launching a major incursion into Angola with a view to preventing the Marxist MPLA from establishing itself as the government of Angola. In this venture the South Africans made common cause with the MPLA's rivals, namely the *Frente*

Nacional de Libertação de Angola (FNLA) and the *União Nacional para a Independência Total de Angola* (UNITA). Such co-operation enabled the South Africans to destroy many of SWAPO's bases in southern Angola, since SWAPO had previously shared certain bases with UNITA and the latter was now willing to reveal the location of these bases to the SADF. However, SWAPO soon recovered from this set-back. Finding itself politically isolated and logistically overstretched, the South African Government pulled its forces out of Angola between January and March 1976. As it did so, SWAPO re-established itself in southern Angola, with the official blessing of the new MPLA Government in Luanda.[25]

The South Africans then attempted to counter SWAPO by blocking off the infiltration routes from southern Angola into northern South West Africa. In May 1976 the SADF launched a major operation, codenamed 'Cobra', designed to eliminate all guerrilla activity in the Ovamboland region. At the same time work began on a scheme to seal the frontier with Angola by creating a 1600 km. (1000 mile) long no-man's-land along the entire length of the northern border. The whole area was cleared of inhabitants and vegetation and a fence erected along the border. A parallel fence was erected one kilometre to the south and the intervening area was declared to be a free-fire zone. Subsequently, between late 1976 and mid-1977, sophisticated electronic warning devices were installed in the no-man's-land, allegedly with Israeli assistance. It was also reported that the SADF sowed the buffer zone with Mexican Sisal, a poisonous plant whose cuts can cause death.

Behind this *cordon sanitaire*, the South Africans attempted to counter SWAPO by four main methods. One was to exploit any political splits within the SWAPO camp. Disillusioned SWAPO leaders were allowed back into South West Africa as free men, in the hope that they might provide a political counterweight to SWAPO. Several former SWAPO leaders, including Andreas Shipanga and Mishake Muyongo, were allowed to return to South West Africa, the former establishing a political party known as SWAPO-Democrats, and the latter, along with other expelled Caprivians, re-establishing the Caprivi African National Union (CANU) outside SWAPO's auspices. More generally, the South Africans offered an amnesty to all SWAPO fighters, in December 1979, in an attempt to encourage desertions from

PLAN. They also released from gaol detained nationalist leaders, including the veteran nationalist Herman Toivo ja Toivo, arrested by the SAP in 1968 and freed in March 1984.

Another method used by the South Africans to counter SWAPO was a civic action programme. As well as pursuing political reforms in order to undermine SWAPO's political appeal, Pretoria also sought to promote its own version of the famous 'hearts and minds' policy. 'Psychological Action' or Psy-Ac officers were posted to each battalion and Psy-Ac manuals were issued to SADF personnel, instructing them how to maintain good relations with the local population. Moreover, the SADF was used as a sort of benign development agency. Troops were detailed to advise and help local people with agricultural and irrigation projects and were also used in a medical and educational role.

By contrast, the SADF also put a considerable effort into countering SWAPO by destroying its in-country organisational efforts. In this respect Pretoria continued and intensified the policies it had introduced before 1974. SWAPO's internal wing was not actually banned, but its political activities were severely circumscribed. The South African Government also extended more of its own security laws to South West Africa, notably the Internal Security Act of 1976, which empowered the authorities to arrest almost anyone suspected of endangering state security or public order, and to detain him/her without charges, bail, trial or counsel. Moreover, the emergency regulations introduced in Ovamboland in 1972 were gradually extended over most of northern South West Africa. In May 1976 Kavangoland and Caprivi (as well as Ovamboland) were declared to be 'security districts' under the control of the SADF. By May 1979, the security district laws had been extended south as far as Windhoek, thus placing some 80 per cent of the population under a form of martial law. In these security districts, the Security Forces were entitled to conduct searches, impose curfews and ban meetings, and possessed wide powers of detention and arrest.

These increasingly harsh security regulations appear to have been accompanied by a growing use of so-called counter-terrorism, especially in Ovamboland. It has been alleged that the Ovambo Home Guards (a police unit organised along the lines of a tribal militia) and Special Constables or Special Police (paramilitary policemen) have harassed and intimidated SWAPO adherents/sympathisers, and that members of *Koevoet* ('Crowbar'), a special COIN unit consisting of white SAP officers and black troops, have assassinated SWAPO supporters as a matter of course. It has also been alleged that the SAP in general has deliberately and systematically used torture in order to extract information from detainees, though the veracity of this charge, like those relating to *Koevoet* and the Ovambo forces, remains open to question.[26]

Attempts to counter SWAPO's military operations within South West Africa centred around standard anti-guerrilla tactics. Troops and policemen were deployed on protection duties, in order to guard against attacks on townships, installations and individuals such as officials, chiefs and headmen. White farms in the so-called 'Murder triangle' around Grootfontein, Tsumeb and Otavi were linked to the SADF's Military Areas Radio Network (MARNET), a measure similar to Rhodesia's Agric-Alert system. Villagers were resettled in protected villages, though apparently not on an extensive scale outside Caprivi and Okavango. The SADF's major effort, however, has gone into preventing SWAPO guerillas from infiltrating beyond the northern tier, that is, the northern border areas. To this end, interdiction or 'search and destroy' operations have been mounted and various special assets, animal, human and material, have been used. Among these assets are Ratel mechanised infantry combat vehicles (MICVs), Eland armoured cars; horses and scrambler motor-cycles for mobility and speed over rough ground; San trackers, who can outrun antelope, and SADF Reconnaissance Commandos (Recces or Recondos), units similar to the Rhodesian Selous Scouts. Use has also been made of helicopters, though because of shortages not on a lavish scale. Instead, SADF ground units have developed Mobile Reaction or Quick Reaction forces, whose role has been to track down and eliminate guerrillas after the Security Forces have learnt, through patrols or through intelligence reports, of a PLAN presence. These Reaction forces consisted of platoon size 'bricks' mounted in mineproofed APCs and maintained at instant readiness in each battalion location.

Follow-up operations against the insurgents have not, however, been confined solely to the territory of South West Africa. As well as pursuing insurgents south of the border, the Security

Forces have also pursued the guerrillas into Zambia, Botswana and Angola. Indeed, during 1976 and 1977, these so-called hot-pursuit operations into neighbouring states became a common occurrence, particularly with regard to Angola. This is not to suggest, though, that incursions into neighbouring states have all been of the hot-pursuit type. On the contrary, the SADF has also used a technique tried and tested by the Israelis and the Rhodesians: pre-emptive ground and air attacks on insurgent bases.

This tactic of hitting the insurgents before they are able to infiltrate into the target territory appears to have been introduced in 1977. In that year, incursions into Angola were gradually extended from ordinary hot pursuit to include carefully planned raids on specific targets and SWAPO, as a result, moved its bases back from the immediate vicinity of the border. The South Africans responded with deep-penetration pre-emptive strikes. The first substantial attack of this kind took place in May 1978, after a build-up of SWAPO strength in southern Angola and an upsurge of guerrilla activity of Ovamboland. Having agreed to Western terms for an international settlement of the South West African question, Pretoria decided to weaken SWAPO by striking at what it believed to be SWAPO's military head-quarters, a planning, training, logistics and communications base located at Cassinga, some 250 km. (155 miles) inside Angola. On 4 May 1978 South African forces launched a day-long raid, which involved air strikes and paradrops as well as a three-pronged ground incursion. According to the South African Government, the SADF destroyed the headquarters and a series of smaller bases nearer the border, and killed hundreds of guerrillas for the loss of only five dead. According to the Angolan Government, however, the victims of the raid — some 1000 people — were refugees, and mostly women and children at that.

From that time on, pre-emptive attacks became a standard feature of South African COIN strategy in South West Africa; attacks were launched both to pre-empt SWAPO offensives and to weaken SWAPO militarily at times when a political settlement looked likely. In March 1979, after SWAPO had conducted a series of attacks against SADF bases and plunged South West Africa into darkness by sabotaging the power lines from the Ruacana power station, the SADF launched three simultaneous raids into Zambia and southern Angola. The South Africans destroyed a number of SWAPO bases without loss, but the results were disappointing because the guerrillas had vacated the bases beforehand. Indeed, during the second half of 1979, SWAPO activity picked up again and having built up its strength to over 8000 men, SWAPO dubbed 1980 a 'year of action'. South Africa responded by launching her biggest combined land and air operation since the Second World War. In a three-week effort during June 1980, her forces overran a sprawling forward operational headquarters codenamed 'Smokeshell', as well as a series of sub-camps, bases and staging posts spread out over 130 sq.km. (50 sq. miles) of southern Angola. According to Pretoria the raid was highly successful. The South Africans claimed to have destroyed 50 tons of equipment and to have captured another 250 tons, including surface-to-air missiles (SAMs), anti-aircraft (AA) guns, armoured personnel carriers (APCs) and recoilless rifles, and to have killed some 360 guerrillas for the loss of only 17 of their own men. They also claimed that SWAPO's operational nerve-centre had been smashed and the guerrillas' efforts set back by months.

Further incursions into Angola followed in 1981, 1982 and 1983. Particularly devastating was Operation 'Protea', a 13-day operation carried out in late August/early September 1981. During the preceding year or so, SWAPO's arsenal had been replenished by the USSR and the movement had regrouped its forces behind the shelter of Angolan/Cuban forces and SAM systems installed by the East Germans. Pretoria's response was to launch yet another ground and air incursion, in which some 4000 men, 41 per cent of whom belonged to SWATF, penetrating over 161 km (100 miles) into Angola. The operation seems to have been a great success. For the loss of only ten dead, the SADF neutralised the air defences, killed around 1000 SWAPO/Cuban/Angolan troops, and seized between 3000 and 4000 tons of equipment, including tanks, APCs, AA guns, SAMs, mortars, minelayers, multiple rocket launchers and millions of rounds of ammunition, collectively valued at over 200 million Rands (£120 million). The South Africans also killed four Russians and captured a Soviet NCO. Operation 'Daisy', a follow-up operation carried out in November 1981, was not so successful. SWAPO units had hastily evacuated their bases before the raiders arrived, though the SADF still claimed to have killed or captured 70 guerrillas and to have destroyed stores of weapons. Two further incursions, in 1982 and 1983 respectively, produced spectacular results. In March 1982 SADF special forces struck at the Cambero Valley, an area of south-western Angola used as a staging post for attacks into Kaokoland and Damaraland, killing 201 insurgents for the loss of three of the raiding party. And in December 1983, after another build-up of SWAPO strength in south-central Angola, the SADF launched a 200 km. (125 mile) deep penetration into southern Angola, codenamed Operation 'Askari', capturing thousands of tons of equipment and killing 500 enemy for the loss of only 21 dead. The 'enemy' included not only SWAPO guerrillas but Angolans and Cubans, killed in a battle at the town of Cuvelai.

As well as launching cross-border raids into Angola the South Africans have also intervened in that country by more covert means. It has been widely reported that one particular SADF unit, 32 or the Buffalo Battalion, has operated in southern Angola on a more or less permanent basis, attacking economic targets, destroying communications and conducting search and destroy sweeps in SWAPO-infested areas. Apparently Buffalo's personnel — ex-FNLA guerrillas officered by white South Africans and former members of the Rhodesian Security Forces — wear unmarked uniforms and use captured Soviet weapons when employed in such a role.[27]

Buffalo personnel have also been used to train the guerrillas of Jonas Savimbi's UNITA, which was often described by the MPLA as being no more than a puppet of the South African Government. Such accusations were not entirely justified, in that UNITA appeared to enjoy a degree of popular support in southeastern Angola. On the other hand, there seems little doubt, despite Pretoria's denials to the contrary, that UNITA and South Africa have collaborated militarily. When the SADF withdrew from Angola in early 1976, UNITA personnel withdrew also, some as far as South West Africa. Subsequently, the SADF took UNITA under its wing, retraining UNITA personnel, supplying them with food and arms, and also, on occasions, handing over to Savimbi's men areas of southern Angola cleared of SWAPO/ MPLA/Cuban forces by SADF raids. From Pretoria's point of view, such collaboration served a most useful purpose. The UNITA connection provided an additional and covert means of intervening in southern Angola, though the precise objective Pretoria had in mind remains conjectural. The South African

Government may have wished to destabilise and topple the MPLA regime (and place UNITA in power either singly or in coalition); to establish a SADF/UNITA buffer zone in southern Angola (sometimes referred to as 'Lebanonisation'); or, alternatively, to exert pressure on the MPLA so as to force the latter to deny bases, and even continued political support, to SWAPO. Whatever Pretoria's objective, or combination of objectives, covert intervention appears to have become another standard feature of its effort to counter the SWAPO threat.

The efficacy of this COIN effort is difficult to assess, given that the conflict has not yet ended and given also that media coverage of the conflict has been restricted. Nevertheless, certain tentative conclusions can be offered. First, the available evidence tends to suggest that the South Africans have successfully blunted SWAPO's military edge. Indeed, the SADF has established a growing mastery over SWAPO in the military sphere, especially since the war was taken to southern Angola. This was a highly dangerous move, inviting not only the risk of UN sanctions but also the wrath of the USSR, which signed a 20-year treaty of mutual friendship and co-operation with Angola in October 1976. However, the results of the South African Government's forward policy were impressive. Successive build-ups of men and material by SWAPO were frustrated by SADF attacks and SWAPO was forced to regroup deeper inside Angola, thus lengthening its infiltration routes and exposing its forces to SADF/UNITA attack long before they reached the border. Moreover, the policy of overt/covert intervention in Angola paid off in other ways. After sustaining over $7 billion worth of damage between 1975 and 1983,[28] the MPLA Government decided to modify its stance on the South West Africa question. By February 1984 the MPLA had entered into a pact with Pretoria: the latter agreed to disengage its forces from southern Angola (the SADF had been occupying parts of southern Angola since 1981), while the Angolans agreed to bar SWAPO from the areas vacated by the South Africans. The eventual outcome of this Lusaka Accord remains to be seen, however.

The South Africans have also made considerable progress in countering SWAPO's military activities *within* South West Africa. While SADF and UNITA operations in southern Angola, and to a lesser extent the establishment of the free-fire zone along the border, reduced infiltration to manageable proportions, the tactics and policies adopted by the South Africans to the south of the border reduced SWAPO's in-country effectiveness. The civic action programme bore fruit in Kaokoland, Kavango-land and Caprivi, while the amnesty policy encouraged, *inter alia*, Caprivian members of SWAPO to defect in 1980; for these reasons, but also because infiltration into these areas has been difficult (Kaokoland has a rugged terrain, while Kavangoland and Caprivi were protected by UNITA's presence to the north), all three regions have been quiet since the late 1970s. The policy of Namibianisation has also been a success. By the early 1980s nearly 30 per cent of all operational forces in South West Africa were local recruits.[29] This greatly reduced the burden on SADF personnel and also gave credence to Pretoria's political init-iatives. Moreover SADF military operations, particularly those of the hot-pursuit variety, have accounted for a comparatively large number of SWAPO guerrillas — according to the South Africans, guerrilla deaths were running at nearly 1500 per year from the late 1970s. By contrast, SADF losses, officially stated to be 50–60 per year during the same period, have not been high enough to undermine South African morale in the field or to engender political disaffection at home. From Pretoria's point of view, therefore, the campaign in South West Africa has been and remains sustainable in military and political terms, and for that matter, in economic terms, even though the war was costing South Africa two million Rands per day by the early 1980s. Thus SWAPO has not been able to make the war prohibitive to the South African Government or to wear down the resolve of the Afrikaner people.

This is not to suggest, though, that the South Africans have completely eradicated the SWAPO threat. SWAPO's military edge may have been blunted, but the Security Forces have not had things all their own way. Skirmishes, sabotage and land-mine explosions have been an everyday occurrence in Ovamboland since the mid-1970s, and on occasions the insurgents have been able to penetrate to the white farming areas south of the northern tier; moreover it cannot be ruled out that SWAPO's military edge may at some stage be resharpened. Furthermore, SWAPO has remained a major threat in the political sphere. Recognising that revolutionary guerrillas cannot usually be defeated by military means alone, the South African Government took various political initiatives in order to counteract SWAPO politically. These measures met with some success, especially outside Ovamboland, but their success was only conditional. Despite the political reforms, and despite the civic action programme, Pretoria has not been able (yet, at least) to check SWAPO politically. SWAPO is said to have the overwhelming support of the Ovambo tribe, whether for ideological or tribal reasons, plus some support outside Ovamboland too. Indeed, many observers, pointing to the fact that the Ovambo now constitute over 50 per cent of the entire population of South West Africa, have predicted that SWAPO would win any UN supervised one-man-one-vote elections hands down. These predictions may or may not prove accurate, but the possibility certainly exists that if such elections were held South Africa might be unable to translate its military ascendancy into political victory. As other COIN campaigns have illustrated, high 'kill ratios' do not necessarily lead to the defeat of the insurgents.

Whatever the eventual outcome of the conflict in South West Africa, that conflict has left its mark on South Africa's entire approach to COIN warfare. Indeed, some of the tactics used in the South West Africa campaign have already been adopted and adapted for use against the ANC, which has been infiltrating guerrillas into South Africa, mostly via Mozambique, since the late 1970s. The ANC's military activities have been of modest proportions to date, but Pretoria's response has been far from complacent. Work has begun on the establishment of a *cordon sanitaire* along the borders with Mozambique and Zimbabwe. Cross-border raids have been launched into Lesotho and Mozambique. And in the case of Mozambique, the South Africans have also, or so it has been alleged, given logistical support to anti-Government guerrillas of the *Resistançia Naçional Moçambicana* (RENAMO or MNR).

These tactics produced the intended results. By early 1984, Mozambique's Marxist regime, which like Angola had entered into treaty relationship with the USSR, was virtually on its knees. Prostrated by MNR insurgency, South African intervention, natural disasters and its own economic mistakes, the Machel Government decided to parley with Pretoria. In March 1984 the Mozambicans agreed to deny base facilities to the ANC in exchange for a South African pledge to cut off support for the MNR. The durability of this Nkomati Accord remains open to question, but the South African Government has achieved at least a breathing space in which to counter the ANC in the political sphere. In this respect, however, Pretoria's approach to countering the ANC may well differ from its approach to countering SWAPO. In South West Africa, institutionalised racial discrimination was abandoned and the territory may well ultimately be seen as negotiable, with an easier defence line being established on the Orange River. The Republic itself, on the other hand, remains non-negotiable, and fundamental political reforms would require the abandonment of the apartheid doctrine. It is partly because of the insurgent threat that P.W. Botha's Government has already introduced a number of radical reforms. However, whether Botha is willing, or will be able, to go further remains for the moment a matter for conjecture.

Notes

1. In June 1968 the UN General Assembly proclaimed that the territory of South West Africa would thenceforth be known as Namibia, a name favoured by the territory's African nationalists. However, although the nationalists and most UN member-states proceeded to use the name Namibia, the South African Government continued to refer to the territory as South West Africa. As this paper is concerned primarily with South African policy, the term South West Africa will be used. No political preference is implied.
2. According to figures cited in *The Apartheid War Machine* (International Defence and Aid Fund, London, 1980), p. 41, the manpower of the SADF in 1960 was as follows: Permanent Force, 11,500; Citizen Force, 2000; Commandos, 48,500; men undergoing national service, 10,000; others (civilian administration, etc.), 6000.
3. Ibid., p. 44.
4. By 1964 the Permanent Force numbered 15,000, the Commandos 51,000 and there were 20,000 men under training. See R.S. Jaster, *South Africa's Narrowing Security Options* (International Institute for Strategic Studies, London, Aldephi Paper no. 159, 1980), p. 12.
5. Ibid.
6. This Act empowered the Minister of Justice to outlaw communist organisations and 'list' (i.e. silence) anyone who in his opinion was likely to further the aims of communism.
7. This Act enabled the Government to declare a state of emergency in part or all of South Africa and rule by proclamation.
8. The PAC, operating in conjunction with the Mozambique nationalist movement *Comité Revoluçionario de Mocambique* (COREMO), tried to infiltrate a small group of its men through Mozambican territory in 1968 but the group was intercepted by the Portuguese at Vila Pery, some 250 km (155 miles) west of Beira. The ANC tried to infiltrate its forces via Rhodesia.
9. Jaster, *South Africa's Narrowing Security Options*, p. 19, citing a report in *The Economist*, 6 March 1969.
10. Jaster, *South Africa's Narrowing Security Options*, p. 19, citing J. Marcum, *The Angolan Revolution* (MIT, Cambridge, Mass., 1978).

11. P.L. Moorcraft and P. McLaughlin, *Chimurenga: The War in Rhodesia, 1965–1980* (Sygma/Collins, Johannesburg, 1982), p. 167. Others have suggested that South Africa had between 2700 and 4000 troops in Rhodesia.

12. In February 1978 the Patriotic Front accused the Rhodesian Government of having recruited 11,200 'mercenaries', including 4500 South Africans, *The Apartheid War Machine*, p. 65.

13. Ibid., pp. 65-6; Moorcraft and McLaughlin, *Chimurenga*, pp. 167-9.

14. For information on Rhodesian COIN tactics, see Chapter 6, above.

15. Even by 1971 SWAPO had only 100 to 150 armed men in South West Africa; K.W. Grundy, *Confrontation and Accommodation in Southern Africa* (University of California Press, Berkeley and Los Angeles, 1973), p. 156.

16. *Sunday Telegraph*, 11 May 1969.

17. The *Guardian*, 15 June 1974. Ten of the dead were policemen.

18. By that time, COIN warfare had become a standard feature of Army training.

19. J.H.P. Serfontein, *Namibia?* (Rex Collings, London, 1976), p. 324.

20. According to C. Legum, *The Western Crisis over Southern Africa* (Africana Publishing Company, London, 1979), p. 184, SWAPO had between 5000 and 10,000 trained fighters by early 1978.

21. *Apartheid's Army in Namibia* (International Defence and Aid Fund, London, 1982), pp. 11, 12 and 3 respectively.

22. W. Gutteridge, *South Africa: Strategy for Survival?* (Conflict Studies, London, Paper no. 131, June 1981), p. 7, for example estimated that there were at that time some 20,000 men under South African control in the border areas.

23. After the mid-1970s South Africa greatly expanded its military manpower. National service had become compulsory for all white youths in 1968 and in 1977 the period of conscription was doubled from 12 to 24 months. The Citizen Force and Commandos were also expanded after the mid-1970s and given a greater role in the defence of the Republic itself, thus freeing troops for service in South West Africa, and non-whites were recruited into the SADF on a voluntary basis. The SAP was also expanded and greater use was made inside the Republic of the Reserve Police and the Police Reserve; for further details, see M. Midlane, 'South Africa', in J.D.P. Keegan (ed.), *World Armies* (Macmillan, London, 1983).

24. On the various equipment bought or manufactured by the South Africans see ibid., and P. Moorcraft, *Africa's Super Power* (Sygma/Collins, Johannesburg, 1981).

25. For accounts of South Africa's intervention in the Angolan conflict, see R. Hallet, 'The South African Intervention in Angola, 1975–6', *African Affairs*, July 1978; A. Gavshon, *Crisis in Africa: Battleground of East and West* (Penguin Books Ltd., Harmondsworth, 1981), pp. 223-57, and I. Greig, *The Communist Challenge to Africa* (The Foreign Affairs Publishing Company, Richmond, Surrey, 1977), pp. 211-36.

26. Allegations such as these have been made by church and humanitarian leaders as well as by SWAPO itself; see *Apartheid's Army in Namibia*, pp. 49-50.

27. Ibid., p. 22.

28. *The Economist*, 16 July 1983.

29. Gutteridge, *South Africa: Strategy for Survival?*, p. 8.

References

Catholic Institute for International Relations and the British Council of Churches, The. *Namibia in the 1980s* (London, 1981)

Davenport, T.R.H. *South Africa: A Modern History* (Macmillan, London, 1978)

Davidson, B., Slovo, J. and Wilkinson, A.R. *Southern Africa: The New Politics of Revolution* (Penguin Books Ltd., Harmondsworth, 1977)

Department of Information and Publicity, SWAPO of Namibia. *To Be Born A Nation: The Liberation Struggle for Namibia* (Zed Press, London, 1981)

Dodd, N.L. 'South African Operations and Deployments in South West Africa/Namibia', *Army Quarterly*, July 1980

Economist, The, 'Destabilising Southern Africa', 16 July 1983

Feit, E. *Urban Revolt in South Africa, 1960-1964* (Northwestern University Press, Chicago, 1971)

Gann, L.H. and Duignan, P. *Why South Africa Will Survive* (Croom Helm, London, 1981)

Geldenhuys, D. and Gutteridge, W. *Instability and Conflict in Southern Africa: South Africa's Role in Regional Security* (Conflict Studies, London, Paper no. 148)

Green, R., Kiljunen, M.-L. and Kiljunen, K. *Namibia The Last Colony* (Longman, Harlow, 1981)

Greig, I. *The Communist Challenge to Africa: An Analysis of Contemporary Soviet, Cuban and Chinese Policies* (The Foreign Affairs Publishing Co. Ltd, Richmond, Surrey, 1977)

Grundy, K.W. *Confrontation and Accommodation in Southern Africa: The Limits of Independence* (University of California Press, Berkeley and Los Angeles, 1973)

Grundy, K.W. 'Namibia in International Politics', *Current History*, March 1982

Gutteridge, W. *South Africa: Strategy for Survival?* (Conflict Studies, London, Paper no. 131)

Henriksen, T.H. 'Namibia: A Comparison with anti-Portuguese Insurgency', *Round Table*, April 1980

International Defence and Aid Fund, The, *Apartheid's Army in Namibia: South Africa's Illegal Military Occupation* (London, 1982)

International Defence and Aid Fund, The, *The Apartheid War Machine: The Strength and Deployment of the South African Armed Forces* (London, 1980)

Jaster, R.S. *South Africa's Narrowing Security Options* (International Institute for Strategic Studies, London, Adelphi Paper no. 159, 1980)

Keegan, J.D.P. (ed.), *World Armies* (Macmillan, London, 1983)

Legum, C. *The Western Crisis over Southern Africa* (Africana Publishing, London, 1979)

Menaul, S. *The Border Wars: South Africa's Response* (Conflict Studies, London, Paper no. 152)

Meredith, M. *The Past as Another Country. Rhodesia: UDI to Zimbabwe* (Pan, London, 1980)

Moorcraft, P.L. *Africa's Super Power* (Sygma/Collins, Johannesburg, 1981)

Moorcraft, P.L. and McLaughlin, P. *Chimurenga: The War in Rhodesia, 1965–1980* (Sygma/Collins, Johannesburg, 1982)

Morris, M. *Southern African Terrorism* (Howard Timmins, Cape Town, 1971)

Plaut, M., Unterhalter, E. and Ward, D. *The Struggle for Southern Africa* (Liberation and War on Want, London, 1981)

Serfontein, J.H.P. *Namibia?* (Rex Collings, London, 1976)

Thompson, L. and Prior, A. *South African Politics* (Yale University Press, New Haven, 1982)

Totemeyer, G. *Namibia Old and New: Traditional and Modern Leaders in Ovamboland* (Cape Hurst and Company, London, 1978)

Totemeyer, G. *South West Africa/Namibia* (Fokus Suid Publishers, Randburg, 1977)

Wellington, J.H. *South West Africa and its Human Issues* (Clarendon Press, Oxford, 1967)

NOTES ON CONTRIBUTORS

Professor Ian F. W. Beckett is Visiting Professor of Military History at the University of Kent. He has previously held chairs at the University of Northampton, the US Marine Corps University, and the University of Luton. His publications include *Modern Insurgencies and Counter-insurgencies*, and *Modern Counter-insurgency*.

Colonel Peter M. Dunn was a serving USAF officer. He is author of *The First Vietnam War* and co-author of *Military Lessons of the Falklands Islands War: Views from the United States*, and *The American War in Vietnam*. He received his doctorate from the University of London.

Major (Ret'd) F. A. Godfrey was formerly Senior Lecturer in War Studies at the Royal Military Academy, Sandhurst. A former Regular soldier, he won the MC in Malaya. He lives in retirement in Norwich.

Dr John Pimlott was Head of the Department of War Studies at the Royal Military Academy, Sandhurst from 1993 until his death in 1997. He had worked at Sandhurst for 24 years. His publications included *Guerrilla Warfare*, *Vietnam: The History and Tactics*, and the *Guinness History of the British Army*.

Dr Francis Toase is Head of the Department of Defence and International Affairs at the Royal Military Academy, Sandhurst. A specialist on Southern Africa, he received his doctorate from the University of Wales in 1984.

INDEX

Abba Siddick 70
Abéché 73
Abrams, Gen. Creighton 88
Adane, Ramdane 62
Aden 19, 23–5, 28–9, 31, 176
Afghan War, Second 2
Afghanistan 9, 11–12
African National Congress (ANC)
   170, 190–1, 194–8, 218–19
African Resistance Movement
   (ARM) 196
Airland Battle doctrine 105
Airwork Services Ltd 36
Akehurst, Brig. John 33, 40–2
Akoot 37
áldeamentos 147, 156, 158
Algeria 4, 8, 11, 17, 46–7, 56, 60–7,
   70, 72–3, 116, 140, 146, 201
Algiers, Battle of 12, 62–4, 66, 116
Alianza Anticommunista Argentine
   (AAA) 122
Alliance for Progress 118
American Civil War 85
American War of Independence 1, 17
Amin, Idi 168
Amritsar 5
Angoche 153
Angola 136, 138–9, 142–5, 147–9,
   150–2, 156, 159, 166, 178, 194, 198,
   201–2, 204, 209–10, 213–16, 218–19
Ankhe 85
Annam 49
Aouzou Strip 71–2
Arab Revolt 5
Arde 120
Arecho, Jose Pacheco 126–8
Argentina 112–13, 116–17, 119, 122–3
Argoud, Col. Antoine 66
Armed forces *see* under individual
   countries
Armed Forces Movement 153
Armée National de Tchad (ANT) 70–1
Army of the Republic of Vietnam
   (ARVN) 83, 90, 94
Arriaga, Kaulza de 141, 149, 155–7
Astiz, Lt. Cmdr. Alfredo 122
Ati, Battle of 72

Bakongo tribe 138
Balante tribe 138
Baluchistan 36
barrages 64
Basmachi Revolt 9
Basutoland *see* Lesotho
Batista, Fulgencio 12, 115–16, 121
Bechuanaland *see* Botswana
Beira 158–9, 164
Beitbridge 199
Belfast 23
Ben Cat 90
Benguela railway 139
Ben M'hidi 64
Ben Suc 90, 92
Berlin 194
Betancourt, Romulo 123
Bié 144
bifurcation 151–2, 177
Bigeard, Col. Marcel 63
Biltine 73
Boer War 2–3, 16–17
Boi Loi woods 90
Bolivia 115, 117, 119, 121
Bonnet, Col. Georges 58
Bordaberry, Juan Maria 123, 128–9,
   131, 133
Borneo 13, 19, 24
Botha P.W. 168, 219
Botswana 164, 167, 169, 180, 196,
   202, 213
Brazil 114, 117, 123
Briggs, Lt. Gen. Sir Harold 23
Britain, Great 164
British Army, 2, 4, 5, 7, 9, 11, 13,
   16–25, 32, 122, 164: Royal
   Artillery 32; Royal Engineers 32;
   Royal Signals 32; 6th Airborne
   Division 12; *see also* Special Air
   Service Regiment
British Army Training Team (BATT)
   32, 36, 39
British Rule, and Rebellion 4
British South Africa Police 170–5,
   183, 186
Brunei 19
Bugeaud, Marshal 4

Bulawayo 185
Bulgaria 140
Burma 2

Cabinda Enclave 150
Cabo Delgado 139, 154, 156
Cabora Bassa 157–8
Cabral, Amilcar 138, 140–1
Caetano, Marcello 142, 152–4
Callwell, Charles 2, 16
Cambero Valley 215
Cameroon 68, 72
Canada 102
Cao Bang 52, 55
Cao Dai 97
Cape Verde 138, 141
Caprivi African National Union
   (CANU) 210
Caprivi Strip 202–3, 205, 211–12, 217
Caracas 123
Cassinga 150, 213
Castro, Fidel 12, 114–17, 124
cavalry 145–6, 175
Centenary 171
Central African Empire/Republic 68
Central Front 99, 105
Central Intelligence Agency (CIA)
   79, 84, 87–8, 96, 98, 120
Chad 46, 68–73
Challe, Gen. Maurice 66
Chan Muong Gorge 57
Cheysson, Claude 73
Chieu Hoi (Open Arms) programme
   95
Chikerema, James 169
Chile 117–18
China, People's Republic of 50–1, 54,
   58, 72, 77, 104, 140, 168–9
Chindits 17
Chinese Civil War 13
Chirau, Jeremiah 166, 170, 184
Chitepo, Herbert 170
Chokwe tribe 138
Chrzanowski, Wojciech 2
Cienfuegos 115
Civil Action Teams (CATs) 11, 33–4,
   37–8, 42
Civil Guard 94
Civil Operations and Rural
   Development Support (CORDS)
   83–4, 94–6
Civilian Irregular Defence Group
   (CIDG) 94, 96–8
Clausewitz, Karl von 1

Cochin China 49
Colombia 104, 115, 118, 120
colons 60, 62, 67
Combined Action Program 89, 98
Combined Operations Headquarters
   (Comops) 171–3, 176
*Comite Revoluçionario de
   Mocambique* (COREMO) 138,
   140–1
command and control 8, 20–1, 56, 79,
   81–4, 131, 142–3, 171–3, 202–3,
   211–12, 217
Commandos Africanos 149
Commandos de Chasse 66
Como Island 144
Companhia Uniao Fabril (CUF) 142,
   153–4
Congo-Brazzaville (Republic of
   Congo) 68, 139, 145
Congo, former Belgian *see* Zaire
*Consejo Anti-communista de
   Guatemala* (CADEG) 122
Constantine 65
cordon sanitaire 11, 179–80, 194–5,
   198, 210, 216, 218
Corvey, Le Miere de 2
Costa Rica 120
Creasy, Maj. Gen. Timothy 39
Cruz, Viriato da 138
Cuba 114–17, 121, 124, 140, 168
Cunene 141
Cuvelai 215
Cyprus 5, 8, 19, 21, 23–5, 116

Damaraland 215
Damavand Line 40
Danang 54, 83, 85
Dan Ve (Self Defence Corps) 94
Dar-es-Salaam 201
Darra Ridge 41
Datton, Tony 181
Davidov, Denis 2
death squads 122–3
Debray, Régis 115
Decker, Karl von 2
Defa 41
Defeating Communist Insurgency 5
Defense Intelligence Agency (DIA)
   88
de Gaulle, Charles 67, 70
de Lattre de Tassigny, Gen. Jean
   52–3, 56
de Lattre Line 53
Dembos region 145

Democratic Turnhalle Alliance (DTA) 206–7
DePuy, Maj. Gen. William 92
Détachement Opérationnel de Protection 63
Dhalqut 41
Dhofar 11, 19, 25–43
Dhofar Charitable Association 27
Dhofar Development Committee (DDC) 33
Dhofar Liberation Front (DLF) 27–9, 34
Diamang 150
Dianas 39
Dien Bien Phu 54, 56–7, 81
'dirty war' 122
Dispositif de Protection Urbaine 63
District Security Assistants (DSAs) 175, 181
Dominican Republic 118
Dong Khe 52

East Germany 129
Ecuador 112, 115, 120
Egypt 140, 201
Ejército de Liberación Nacional (ELN) 117
Ejército Revolucionario del Pueblo (ERP) 117, 122
El Salvador 104, 114, 119, 124
Elbrick, Burke 123
Emmerich, Andreas 1
EOKA (National Organisation of Cypriot Fighters) 116
Evian Agreement 67
Ewald, Johann von 1

Falkland Islands 112–13
Farabundo Marti National Liberation Front 120
Fatima 141
Faya Largeau 72–3
Fidelistas 12, 115
Filhol Plantation 90
Fire Force concept 10, 177–8
Firqat 11, 35–8, 40
Flechas 149, 158, 176
Fletcher, Brig. Jack 39
foco theory 115–17, 124
'Football War' 114
Forces Armée du Nord (FAN) 72
Forces Armées Nationales Tchadiennes (FANT) 72–3
Force d'Assistance Rapide 73

force d'intervention 68, 73
Fort Bragg 118, 120
Fort Gulick 118–19
Fort Lamy *see* N'Djamena
Franco-Prussian War 2–3
francs tireurs 2–3
French Army 4, 8–9, 11–12, 46–7, 51, 58, 62, 66–7, 71: 10th Colonial Parachute Division 12; parachutists 52–4, 57, 63–4, 66–8
French Foreign Legion 53–4, 66, 70–2
French Indochina 4, 6, 10, 17, 46–7, 49–58, 60, 64–5
French Marines 55, 66, 68, 71–2
French Revolutionary War 1
Frente de Libertação de Moçambique (FRELIMO) 138–41, 154–9
Frente Nacional de Libertação de Angola (FNLA) 138–40, 144, 209–10
Frente para a Libertação e Independência de Guine (FLING) 138
Frepalina 117
Front de Libération Nationale (FLN) 60, 62–7, 116
Front de Libération Nationale de Tchad (Frolinat) 70–2
Front for the Liberation of Zimbabwe (FROLIZI) 169
Fuerzas Armadas de Liberación Nacional (FALN) 123
Fula tribe 139, 149

Gabon 68
Gallieni, Joseph 4
Galula, David 7
Gavin, James 85
General Law Amendment Acts 196
Geneva 164, 166, 169
Giap, Vo Nguyen 5, 51–5
Girling, J.L.S. 7
Giscard d'Estaing, Valérie 71
Goa 136, 154
Godard, Col. Yves 63
Gomes, Gen. Costa 152
Goukoni Queddei 71–3
Gouvernement d'Union Nationale de Transition (GUNT) 72–3
Graham, Brig. John 30
Granma 115
Greece 13, 79
Grenada 118

Grey's Scouts 175
Grootfontein 205, 209, 212
Groupements de Commandos Mixtes Aéroportés (GCMA) 56, 58
Groupements Mixtes d'Intervention (GMI) 56, 58
Grupos Espeçiais (GE) 149, 152, 157
Grupos Espeçiais de Paraqedist (GEP) 149, 152, 157
Guard Force 175, 181, 184
Guatemala 104, 115–16, 118, 120–2
guerre révolutionnaire 4, 6, 10, 19, 46, 58–60, 62–7, 73
Guevara, Ernesto 'Ché' 5, 115, 119, 121
Guillen, Abraham 117
Guinea, Republic of 139, 140, 148
Guinea-Bissau *see* Portuguese Guinea
Gwynn Sir Charles 4–5

Haiphong 50, 53
Hajar 26
Haluf 37
Hamlet Evaluation System (HES) 94–5
Hammer Line 40
Hanoi 50, 52, 54
Harding, Field Marshal Lord 8
Harkis 66
Harvey, Col. Michael 29
Hauf 28
Havanja 115
hearts and minds, winning (WHAM) 3, 11, 22, 32, 35, 65, 120–2, 148–9, 156–7, 181–2, 211, 217–18
helicopters 10, 86–7, 144, 146, 178, 207
Herreros 3, 147, 200
Hickman, John 171, 173
Hilsman, Roger 93
Hissène Habré 71–2
Ho Bo woods 90
Hoa Binh 53
Hoa Hao 97
Ho Chi Minh 50–1
Ho Chi Minh Trail 77
Hogard, Commander J. 59
Hoi Chanhs ('ralliers to South Vietnam') 89
Honduras 104, 114, 120
Hornbeam Line 40
hot pursuit 9, 139–40, 167, 212–13, 217

Hottentots 3
Hukbalahap 6–7, 80, 92
Hunt Committee 21

Iman of Oman 26–7
Imperial Iranian Battle Group (IIBG) 36, 39–42
Imperial Policing 4
India 9, 136
intelligence 8, 21, 56, 65, 88–90, 98–9, 132, 143, 172–3, 195, 202
Internal Security Act 211
International Court of Justice 200
Iraq 9, 27, 34
Ireland 5
Irish Republican Army (IRA) 17
'Iron Triangle' 90, 92
Israel 9, 142, 176, 210
Izki 30

Jackson, Geoffrey 128
'Jasmines' 39
Jeapes, Col. Tony 32–3, 43
Jebel 25–41 *passim*
Jibjat 38
Johannesburg 191
Johnson, Gen. Harold 85
Johnson, Lyndon B. 84, 93
Joint Chiefs of Staff (JCS) 82
joint action companies 98–9
Joint Operations Centres 171, 179
Jordan 37, 41–2

Kabylia 62, 66
Kaffir factor ('K' factor) 186
Kaffirs 2
Kaokoland 215, 217
Karanga tribe 174
Kassangula 167
Kaunda, Kenneth 164, 166–7
Kavandame, Lazaro 157
Kavangoland 202–3, 205, 211–12, 217
Kealy, Capt. Mike 39
Keegan, Maj-Gen. George 88
Kennedy, John F. 93, 96, 118, 120
Kenya 13, 19, 22–4, 183
Khrushchev, Nikita 118
Kikuyu 22
kill ratios 179, 218
Kissinger, Henry 164
Kitson, Gen. Sir Frank 8, 24
Komer, Robert 84
Korean War 79–80, 83, 128, 196
Kurds 9

Kuwait 176

Lacheroy, Col. Charles 6, 58, 65
Lacoste, Robert 63
Lake Chad 72
Lamarca, Carlos 114
Lancaster House Conference 166
Lang Son 52
Lansdale, Maj-Gen. Edward 80
Laos 49–50, 53–5, 79, 83
Latin America 7, 11, 12–13, 112–33
    passim
Lawrence, T.E. 17
League of Dhofari Soldiers 27
League of Nations 199
Lebanisation 216
Leopard Line 39
Lesotho 194, 218
Lewis, Col. Anthony 28
Libya 70–3
local forces 11, 22–3, 35–6, 56–7, 66,
    94–8, 149, 152, 157, 174, 183, 208,
    211–12, 215–16
Londonderry 23
Long Range Desert Group 17
Lourenço Marques 152, 157
Luanda 138
Lusaka 167, 170, 216
Lyautey Herbert 4

McChristian, Brig-Gen. J.A. 89
McCuen, John 6, 7
Machel, Samora 139, 166, 168, 218
McNamara, Robert 84, 101
McNamara Line 11, 146
Macua 139
Madagascar 4, 47
Magsaysay, Ramon 7, 80, 92
Makonde 138–9, 155, 157
Malawi 139–40, 157–8, 163
Malaya 6–7, 10–11, 13, 17, 19, 20–4,
    56, 92, 99, 145, 176, 180
Malayan People's Anti-Japanese
    Army (MPAJA) 17
Malayan Races Liberation Army 17
Malloum, Gen. 70–2
Mandela Nelson, 196
Mandinka tribe 139
Mao Blanca (MANO) 122
Mao Khe 53
Mao Tse-tung 1, 5, 51, 58–9, 77, 80,
    85, 92, 115, 124, 169, 176
Maoris 2
Marighela, Carlos 5, 117, 123

'Mashford's Militia' 174
Mashonaland 163
Massu, Gen. Jacques 63–4, 66–7
Matabeleland 163
Mau Mau *see* Kenya
Mauritania 68
Mazoe river 179
Mbundu 138–9
Mecca 149
Medinat Al Haq ('White City') 38
Mekong Delta 87, 93
Mendés-France, Pierre 55
Mexico 104, 114
Meyer, Gen. E.C. 106
Midway Road 26–8, 37, 39, 40
Military Areas Radio Network
    (MARNET) 212
Military Assistance Command,
    Vietnam (MACV) 82–4, 88–9
Mirbat 25, 28–30, 38–40
Miristas 117
Mitrione, Dan 119, 128
Mitterand, François 73
Moamar al Gaddafi, Col. 71–2
Mobile Riverine Force 87
Mondlane, Edouardo 139–40, 155
Monroe Doctrine 118
montagnards 96–7
Montevideo 125, 127–8
Montoneros 122
Morice Line 11, 64, 146
Morocco 4, 62, 64–5
Moussoro 73
*Mouvement National pour la
    Révolution Culturelle et Sociale* 71
*Movimento Popular de Libertação de
    Angola* (MPLA) 138–40, 144–5,
    204, 209–10, 216
Moxico 144, 147
Mozambique 136, 138–9, 141–7,
    149–50, 152–9, 164, 166–9, 178–80,
    185–6, 194, 198, 218
Mpacha 202
Mudhai 28
Mueda 138, 155
Mugabe, Robert 170, 174
Mughsayl 38–40
Muhammad Ahmad al-Ghassani 29
Multi-Party Conference (MPC) 207
'Murder triangle' 212
Murrupa, Miguel 157
Musandam peninsula 25
Musergedzi river 179
Muyongo, Mishake 210

Muzorewa, Abel 166, 169–70, 172, 174, 184, 199
My Lai 100

Namibia *see* South West Africa
Namibianisation 207–8, 217
Nampula 157
Nangade 156
Na San 53–4
National Democratic Front for the Liberation of Oman and the Arabian Gulf (NDFLOAG) 30
National Security Study Memorandum (NSSM) 96
Navarre, Gen. Henri 54, 56
Ndebele tribe 169–70, *see also* Matabele
N'Djamena 70–3
Negd 26
Neto, Angostinho 138, 141
New Chimoio 166, 168, 178
Nha Trang 97
Nhari Rebellion 170
Niassa 139, 155–6
Nicaragua 104, 119–20, 124
Nicaraguan Democratic Front (FDN) 120
Niger 68
Nigeria 70, 72, 140
Nizwa 30
Nkomati Accords 218–9
Nkomo, Joshua 163–4, 167–9
North Atlantic Treaty Organisation (NATO) 141, 149–50
Northern Ireland 19, 21, 24–5, 31
North Korea 79
North Vietnam 77, 79, 81, 83, 85, 92
North Vietnamese Army (NVA) 88, 95
Nueva Organizacion Anti-communista (NOA) 122
Nujoma, Sam 201
Nyanja tribe 138, 155
Nyasaland *see* Malawi

Odendaal Plan 200, 203
Okahandja 208
Okavango 208, 212
Okinawa 97
Oman 11, 13, 25–43 *passim*, 184
Omani Army *see* Sultan's Armed Forces
Ondangua 202
Ongulumbashe 201

Operations: 'Apio' 155; 'Askari' 215; 'Atlante' 54; 'Attila' 144; 'Castor' 54; 'Cedar Falls' 89–90, 92; 'Cobra' 210; 'Daisy' 214; 'Dharab' 40; 'Eland' 167; 'Favour' 184; 'Frontier' 156; 'Garotte' 155; 'Gordian Knot' 155; 'Hadaf' 41; 'Himaar' 40; 'Hornet' 37; 'Jaguar' 38; 'Leopard' 38; 'Lorraine' 53, 57; 'Manila Interface' 181; 'Miracle' 166; 'Motorman' 23; 'Overload' 180; 'Panther' 38; 'Protea' 214; 'Rainbow' 28; 'Saeed' 41; 'Simba' 38; 'Turkey' 182
Oppenheim, Senior Assistant Commissioner 183
Oran 66
Orange river 219
Organisation of African Unity (OAU) 141, 168–9
organigramme 63
Organisation Armée Secrète (OAS) 67
Oriente 115
Osorio, Col. Arana 121–2
Ossewa Brandwag ('Ox-Wagon Sentinel') 194
Otavi 205, 212
Ouarsenis mountains 66
Ovambo 147, 201–4, 208, 218
Ovambo Home Guard 211–12
Ovambo Special Constables 212
Ovamboland 201–5, 210–13, 217–18
Ovamboland People's Congress 201
Ovamboland People's Organisation 201
Ovimbundu tribe 138

Paardeberg 3
pacification (French) 48, 49, 52–3, 57, 65, 70
Palestine 5, 12, 17, 19
Palestine Police 21
Pan-Africanist Congress (PAC) 141, 191, 194–8
Panama Canal 118–19
Pando 127
Paraguay 117
Paris 51, 55–6
Paris Peace Accord 100
Partido Africano da Independência de Guiné e Cabo Verde (PAIGC) 138, 140–1, 144, 146
Pearce Commission 164, 174

Peking 51
Pentagon 82
People's Democratic Republic of
  Yemen (PDRY) 25, 29, 34, 38–41
People's Liberation Army of
  Namibia (PLAN) 201–19 *passim*
People's Self-Defence Forces (PSDF)
  95
Perez, Col. Barrera 121
Perkins. Maj-Gen. Kenneth 41
Peru 104, 115–16, 118, 121, 124
Peterson, Tommy 183
Phat Diem 53
Philippeville 62
Philippine Civic Action Group 83
Philippines 6–7, 12–13, 80, 92
'Phoenix' programme 95–6
Phu Doan 53
Phu Hoa Dong 90
Phu Tho 53
pied noir see colons
Plain of Jars 53
Plan Lazo 121
Plan Piloto 121
Pleiku 85
police 8, 21, 126–7, 130–1, 140, 143,
  158–9, 170–3, 181, 190–219 *passim*
politicisation of armed forces 12,
  66–7, 123, 129–33, 152–4, 177
Pope Paul VI 141
Popski's Private Army 17
Popular Forces (PF) 94–5, 98
Popular Front for the Liberation of
  the Occupied Arabian Gulf
  (PFLOAG) 29–40 *passim*
Popular Front for the Liberation of
  Oman (PFLO) 40–1
Poqo 191, 194–8
Port Elizabeth 191
Portugal 8–9, 136–162 *passim*, 206
Portugal and the Future 153–4
Portuguese Army 8–9, 11–12,
  136–62 *passim*
Portuguese Guinea 136, 138–40,
  142–54, 159
Powell, Lt. 'Spike' 183
Project 'Cohort' 106
Project '100,000' 102
propaganda 12, 21, 30, 34–5, 42, 55,
  140–1, 152, 157, 181–2
protected villages *see* resettlement
pseudo-operations 12, 24, 149, 183–4
Psychological Action and
  Information Service 65

Public Safety Act 196
Purdon, Brig. Corin 29
Pushtay, John 6–7

Qaboos bin Said, Sultan 31–42
  passim
Q-cars 179
Quadrillage 65

Radfan 19
Rakhuyt 25, 30, 40–1
Ralil 212
ratissage 66
Raysut 42
Rebelo Decrees 153
'Red Line' 73
Red river 51–3, 57
Regional Forces (RF) 94–5
Reid-Daly, Ron 172–3, 176, 183
Réserve Générale 66
resettlement 11, 22–3, 57, 59, 65,
  93–4, 147–8, 175, 177, 184–5, 212
Resistançia Naçional Moçambicana
  (RENAMO) 218
Revolutionary Development (RD)
  programme 94
Rhade tribe 97
Rheault, Col. Robert 97
Rhodes, Cecil 163
Rhodesia 8–12, 141, 146, 158,
  163–87, 190, 194, 198–9
Rhodesia Regiment 173
Rhodesian African Rifles (RAR)
  173–6
Rhodesian Army 141, 158, 163–83
Rhodesian Defence Regiment 175
Rhodesian Light Infantry (RLI) 175,
  177
Rivonia 196
Roberto Holden 138
Romer, Robert 94
Rovuma river 144, 154
Royal Air Force Regiment 32
Royal Ulster Constabulary (RUC) 21
Ruacana 209, 213
Rundu 202
Rural Industrial Development
  Agency 11

Said bin Taimur, Sultan 26–7, 29, 30,
  36–7
Saigon 50, 55, 82
Saigon river 90
Salal 73

Salalah 25–42 *passim*
Salan, Gen. Raoul 53–6, 67
Salazar, Antonio de 136, 142–3, 152–3
Salim Mubarak 34–5
Salisbury 175, 185
San 208, 213
Sansom, Dr. Robert 88
San Vicente 120
Sandinistas 120
Santos, Marcelino dos 141
Sarfait 38, 41
Saudi Arabia 25, 27–8
Savimbi, Jonas, 138, 215
Savoury, Lt. Alan 183
Sayyid Tariq bin Taimur 31
Schultz, Arnaldo 143, 147–8
Search and destroy 85, 93, 212
Security Force Auxiliaries (SFAs) 172, 175, 184, 186
Selous Scouts 172–3, 175–8, 183–4, 212
Sendero Luminoso 124
Senegal 68, 139, 148, 154
Seroi Pra'ak 11
Seven Pillars of Wisdom 17
Shackleton, Capt. Ron 97
Sharpeville 191
Sheppard Group 181
Sheppard, Ian 181
Sherishitti 37, 40–1
Shipanga, Andreas, 210
Shona tribe 169–70, 181–2 *see also* Mashona
Sierra Maestros 115
Sithole, Ndabaningi 166, 169, 172, 174, 184
Small Wars 2, 16
Smith, Ian 164, 166, 168, 171–2
'smokeshell' 212
Smuts, Jan 194
Soames, Lord 166
Somoza, Anastias 120
Souk-Ahras 65
South Africa, Republic of 9, 11, 141, 158, 164, 167–8, 175, 186, 190–219
South African Communist Party (SACP) 191, 194–8, 218–19
South African Defence Force (SADF) 190–219 *passim*: Buffalo Battalion 11, 215; Cape Corps 207; Commandos 194, 207–8; Citizen Force 194, 207; Koevoet 212; Permanent Force (PF) 194, 207;

Reconnaissance Commandos 212; 21 Infantry Battalion 207
South African Native Affairs Department 200
South African Police (SAP) 190–219 *passim*
South Georgia 122
South Korea 79
South Vietnam 77–107 *passim*
South West Africa (Namibia) 3, 11, 190–1, 199–219
South West Africa People's Organisation (SWAPO) 200–19 *passim*
South West Africa People's Organisation-Democrats (SWAPO (D)) 210
South West Africa Police (SWAP) 208–9
South West Africa Territory Force (SWATF) 11, 208–9, 214
South West African National Union (SWANU) 200–1
Soviet Union 5, 9–11, 34, 104, 140, 168–9, 201, 214, 216, 218
Spain, 2
Special Administrative Section (SAS) 65
Special Air Service (SAS): British 17, 30–41 *passim*; Rhodesian 172, 175–6, 178
Special Night Squads 17
Special Operations Executive (SOE) 17
Spinola, Antonio de 143, 147–9, 154
spirit mediums 182
Stalin, Joseph 9
Stimson, H.J. 4
Stolzman, K.B. 2
Strait of Hormuz 25, 31
Strategic Hamlets Program (SHP) 93–4
Sudan 70
Sudh 30, 35, 37
Sultan's Armed Forces (SAF) 28–41 *passim*: Desert Regiment (DR) 36, 38–9; Dhofar Brigade 36–7, 39–40; Frontier Force (FF) 36, 40–1; Jebel Regiment (JR) 36, 39–40; Kateebat Janoobiya (Southern Regiment) (KJ) 36; Muscat Regiment (MR) 27–8, 36, 40–1; Northern Frontier Regiment (NFR) 27–9, 36–7, 39; Oman

Artillery 36–7; Oman Gendarmerie
36; SAF Armoured Car Squadron
36
Sultan of Oman's Air Force (SOAF)
36–7, 39, 41
Sultan of Oman's Navy (SON) 36
Suppression of Communism Act
208–9, 214
Swaziland 194
Sweden 102

Taber, Robert 6, 116
Tache d'huile 4
T'ai forces 54, 56
Tangwena tribe 183
Tanzania 139–40, 144, 154, 201
Taqa 25, 28, 30, 40
Tauri Attair 38
Taylor, Maxwell 82, 84
Templer, Gen. Sir Gerald 8, 20, 22
Terrorism Act 203
Terrorist Victims Relief Fund 186
Tet Offensive 87–8, 95
Tete 157–9
Than Dien forest 90
Thi Tinh river 90
Thompson, Sir Robert 5, 10, 20,
81–2, 93
Thumrait 26, 37
Tibesti 70, 72–3
Toivo ja Toivo, Herman 211
Tombalbaye, François 70–1
Tonkin 49–50
Tourane *see* Danang
Transvaal 198
Tribal Trust Lands (TTLs) 176–7,
180–1, 183–5
Trinquier, Col. Roger 6, 58, 63, 105
Tsumeb 205, 212
Tukhachevski, Semyon 9
Tunisia 62, 64, 67
Tupamaros 117, 119, 121, 124–33
passim
Turnhalle 205–6
26 July Movement 115
Tyrol 2

Uganda 168
Umkhonto we Sizwe 191, 194–8
Umtali 185
União das Populações de Angola
(UPA) 138
*União Nacional para a Independência
Total de Angola* (UNITA) 138–40,

210, 215–16
United African National Council
(UANC) 169, 170, 184
United Arab Emirates (UAE) 25
United Nations 140, 168, 194, 200,
203–6, 216, 218
United States of America 6, 72, 77,
118–21, 123, 141, 164
United States Agency for
International Development (AID)
79, 84, 119
United States Air Force 79, 81, 83,
87
United States Army 3, 7, 9–10,
77–107 *passim*, 118–19, 133: 11th
Armored Cavalry Regiment 90;
5th Special Forces Group 97; 1st
Air Cavalry Division 85–6; 1st
Infantry Division 90, 92; 196th
Light Infantry Brigade 90; 173rd
Airborne Brigade 90; 25th Infantry
Division 90
United States Congress 82
United States Marines 79, 81, 83, 85,
98–9
United States Navy 81
United States 7th Fleet 83
United States Special Forces 79,
84–5, 94, 96–8, 106, 118–19
United States State Department 79,
82, 84, 88
urban guerrilla warfare 116–17, 124
Uruguay 116–17, 119, 121, 123–33
passim
Usulutan 120

Valentini, George Wilhelm von 1
Van Fleet Mission 79
Vendée 2
Venezuela 116, 118, 123–4
Venter, A.J. 151
Verwoerd, Hendrik 197
Victims of Terrorism (Compensation)
Bill 186
Victoria Falls 164
Viet Bac 51
Viet Cong (VC) 77–98 *passim*
Viet Cong Infrastructure (VCI) 96,
98
Viet Cong 272nd Regiment 90
Vietnam 7, 9–13, 49–58 *passim*, 62,
77, 80–107 *passim* 142, 144–5, 151
*Viet Nam Doc Lap Dong Minh
Hoi* (Viet Minh) 50–8 *passim*

Vietnamese Communist Party 49
Vietnamization 88, 95
Vinh Yen 51
Vorster, John 164, 167–8, 195

Wadi Ashoq 40
Wadi Darbat 38
Wadi Sayq 41
Walls, Lt-Gen. Peter 171–2, 176
Walvis Bay 209
War of the Flea 6
Washington 81–2, 84, 88, 96, 102
Watts, Lt-Col. John 30, 32
'Week of the Barricades' 67
Western Sahara 68
Westmoreland, Gen. William 82–5, 88, 94, 99–100, 103
White House 82–4
Wickham, Gen. 106
Wilson, Harold 164
Windhoek 208–9, 211
Wingate, Orde 17
Wiriyamu 152
World War: First 136, 199; Second 1, 5, 17, 47, 50, 79, 136, 194

Yacef, Saadi 63–4
Yao tribe 155
Young, Sir Arthur 21

Zacapa 121
Zaire 68, 72, 139, 163
Zambezi river 169–70, 180, 183
Zambezia 159
Zambia 139–40, 145, 158, 163–4, 166–7, 169, 170, 178, 180, 185, 198, 201–2, 213
Zimbabwe *see* Rhodesia
Zimbabwe African National Liberation Army (ZANLA) 166–71, 179, 183, 186
Zimbabwe African National Union (ZANU) 158, 168–9, 184
Zimbabwe African People's Union (ZAPU) 163–4, 168–70
Zimbabwe People's Revolutionary Army (ZIPRA) 167, 169, 183, 186
Zulus 2